Martin Wendt

Cosmological variation of the proton-to-electron mass ratio

Martin Wendt

Cosmological variation of the proton-to-electron mass ratio

Probing cosmological variation of the proton-to-electron mass ratio by means of quasar absorption spectra

Südwestdeutscher Verlag für Hochschulschriften

Imprint
Any brand names and product names mentioned in this book are subject to trademark, brand or patent protection and are trademarks or registered trademarks of their respective holders. The use of brand names, product names, common names, trade names, product descriptions etc. even without a particular marking in this work is in no way to be construed to mean that such names may be regarded as unrestricted in respect of trademark and brand protection legislation and could thus be used by anyone.

Publisher:
Südwestdeutscher Verlag für Hochschulschriften
is a trademark of
Dodo Books Indian Ocean Ltd., member of the OmniScriptum S.R.L Publishing group
str. A.Russo 15, of. 61, Chisinau-2068, Republic of Moldova Europe
Printed at: see last page
ISBN: 978-3-8381-2077-5

Zugl. / Approved by: Hamburg, Uni, Diss., 2010

Copyright © Martin Wendt
Copyright © 2010 Dodo Books Indian Ocean Ltd., member of the OmniScriptum S.R.L Publishing group

Contents

1 **Introduction** — 3

2 **Background** — 6
  2.1 Cosmology of varying fundamental constants — 6
    2.1.1 Accessible constants — 10
  2.2 Observables — 12
    2.2.1 Proton-to-electron mass ratio μ — 12
    2.2.2 Intergalactic H_2 — 12
    2.2.3 Laboratory wavelengths — 15
    2.2.4 Sensitivity coefficients K_i — 16
    2.2.5 Status quo for $\Delta\mu/\mu$ — 19

3 **Analysis I** — 22
  3.1 Data — 22
    3.1.1 QSO 0347-383 — 22
    3.1.2 Reduction — 24
  3.2 Preprocessing of data — 26
    3.2.1 Conditioning of flux — 26
    3.2.2 Correction for individual shifts — 26
    3.2.3 Selection of H_2 lines — 29
  3.3 Fitting — 36
    3.3.1 Simultaneous fit vs. co-added fit — 38
    3.3.2 Continuum handling — 38

4 **Results I** — 41
  4.1 Determination of $\Delta\mu/\mu$ — 41
  4.2 Result via discrete line pairs — 46

5 **Error Analysis I** — 49
  5.1 Quality of fit — 49
  5.2 Standard Error — 51
  5.3 Uncertainties in the sensitivity coefficients — 52
  5.4 Influence analysis of data preprocessing — 54

5.5		Rotational Levels - medium dependent	55
5.6		Vibrational Levels - energy dependent	57
5.7		Electronic levels - Lyman, Werner bands	59
5.8		Accuracy of line fits	61

6 Analysis II 65
6.1 Data 65
6.1.1 2009 observations 65
6.1.2 Reduction 68
6.2 Preprocessing of data 71
6.2.1 Correction for individual shifts 71
6.2.2 Selection of H_2 lines 71
6.2.3 Resolution 74

7 Results II 77
7.1 Determination of $\Delta\mu/\mu$ 77
7.2 Result via discrete line pairs 81

8 Error Analysis II 83
8.1 Impact of wavelength calibration issues 83
8.2 Test for correlation of redshift and photon energy 85
8.3 Variability of QSO 0347-383 86
8.4 Calibration and positioning errors 88

9 Conclusions 91
9.1 Inference on cosmology 91
9.2 Outlook 94

Overview of written Programs   96

List of Figures   99

List of Tables   101

Acknowledgements   103

References   105

1 Introduction

The Standard Model of particle physics (SMPP) is very successful and its predictions are tested to high precision in laboratories around the world. SMPP needs several dimensionless fundamental constants, such as coupling constants and mass ratios, whose values cannot be predicted and must be established through experiment (Fritzsch 2009). Our confidence in their constancy stems from laboratory experiments over human time-scales but variations might have occurred over the 14 billion-year history of the Universe while remaining undetectably small today.

The possible variation of the fundamental constants of nature is currently a very popular research topic and has a long history (Dirac 1937; Gamow 1967) Theories unifying gravity and other interactions suggest the possibility of spatial and temporal variation of physical "constants" in the Universe (see, e.g. Marciano 1984; Uzan 2003). Current interest is high because in superstring theories – which have additional dimensions compactified on tiny scales – any variation of the size of the extra dimensions results in changes in the 3-dimensional coupling constants. At present no mechanism for keeping the spatial scale static has been found (e.g., our three "large" spatial dimensions increase in size). Moreover, there exists a mechanism for making all coupling constants and masses of elementary particles both space and time dependent, and influenced by local circumstances (see, e.g., Uzan 2003). The variation of coupling constants can be non-monotonic (e.g., damped oscillations). Indeed, in theoretical models seeking to unify the four forces of nature, the coupling constants vary naturally on cosmological scales. The proton-to-electron mass ratio, $\mu = m_p/m_e$ has been the subject of numerous studies. The mass ratio is sensitive primarily to the quantum chromodynamic scale. The Λ_{QCD} scale should vary considerably faster than that of quantum electrodynamics Λ_{QED}. As a consequence, the secular change in the proton-to-electron mass ratio, if any, should be larger than that of the fine structure constant. This makes μ a very interesting target to search for possible cosmological variations of the fundamental constants. The present value of the proton-to-electron mass ratio is $\mu = 1836.15267261(85)$ (Mohr and Taylor 2005). Laboratory experiments by comparing the rates between clocks based on hyperfine transitions in atoms with a different dependence on μ restrict the time-dependence of μ at the level of $(\dot{\mu}/\mu)_{t_0} = (1.6 \pm 1.7) \times 10^{-15}$ yr^{-1} (Blatt et al. 2008).

A probe of the variation of μ is obtained by comparing rotational versus vibrational modes of molecules as first suggested by Thompson (1975). The method is based on the fact that the wavelengths of vibro-rotational lines of molecules depend on the reduced mass, M, of the molecule. The energy difference between two consecutive levels of the rotational spectrum of a diatomic molecule scale as M, whereas the energy difference between two adjacent levels of the vibrational spectrum is proportional to $M^{-1/2}$:

$$\nu \sim c_{\text{elec}} + \frac{c_{\text{vib}}}{\mu^{1/2}} + \frac{c_{\text{rot}}}{\mu}. \tag{1.1}$$

Consequently, by studying the Lyman and Werner transitions of molecular hydrogen we may obtain information about a change in μ. The observed wavelength λ of any given line in an absorption system at the redshift z differs from the local rest-frame wavelength λ_0 of the same line in the laboratory according to the relation

$$\lambda = \lambda_0(1+z)\left(1 + K\frac{\Delta\mu}{\mu}\right) \tag{1.2}$$

where K is the sensitivity coefficient computed theoretically for the Lyman and Werner bands of the H_2 molecule. Using this expression, the cosmological redshift of a line can be distinguished from the shift due to a variation of μ.

This method was used to obtain upper bounds on the secular variation of the proton-to-electron mass ratio from observations of distant absorption systems in the spectra of quasars at several redshifts. The quasar absorption system towards QSO 0347-383 was first studied using high-resolution spectra obtained with the Very Large Telescope/Ultraviolet-Visual Echelle Spectrograph (VLT/UVES). A first stringent bound was derived at $(-1.8 \pm 3.8) \times 10^{-5}$ (Levshakov et al. 2002). Subsequent measures of the quasar absorption systems of QSO 0347-382 and QSO 1232+082 provided hints for a variation at 3.5 σ (Reinhold et al. 2006; Ivanchik et al. 2005; Ubachs et al. 2007):

$$\Delta\mu/\mu = (2.4 \pm 0.6) \times 10^{-5}. \tag{1.3}$$

The new analysis used additional high-resolution spectra and updated laboratory data of the energy levels and of the rest frame wavelengths of the H_2 molecule. However, more recently King et al. (2008); Wendt and Reimers (2008) and Thompson et al. (2009a) re-evaluated data of the same system and report a result in agreement with no variation. The most stringent limits on $\Delta\mu/\mu$ have been reported at $\Delta\mu/\mu = (2.6 \pm 3.0_{\text{stat}}) \times 10^{-6}$ from the combination of three H_2 systems (King et al. 2008) and a fourth one has provided $(+5.6 \pm 5.5_{\text{stat}} \pm 2.7_{\text{sys}}) \times 10^{-6}$ (Malec et al. 2010).

In this work, the same data of QSO 0347-383 that led in parts to the above mentioned results is analyzed in combination with supplemental observations carried out independently at the same time by another group. This utilization of previously overlooked data enables an improved analysis of the systematics involved and increases the total signal-to-noise ratio notably.

The corresponding analysis and its detailed error handling are described in chapter **Analysis I** on page 22 and following pages. The present analysis is motivated on one side by the use of a new data set available in the ESO data archive and previously overlooked and on the other side by numerous findings of different groups that partially are in disagreement with each other. A large part of these discrepancies reflect the different methods of handling systematic errors.

Evidently systematics are not yet under control or fully understood. This work emphasizes the importance to take these errors, in particular calibration issues, into account and put forward some measures adapted to the problem.

The second part of this thesis deals with the latest observations in the field of variation in fundamental physical constants. Recorded at UVES/VLT in September 2009, the telescope setup during observations and the following data reduction were carried out with the needs for highest precision in mind. The obtained data is of high-quality and its detailed analysis yields the most stringent constraint on $\Delta\mu$ for a single absorber. The methods involved in the determination of μ and the refinements in the data analysis are illustrated in chapter **Analysis II** on page 65 and following pages.

The bounds on the variation of μ are generally obtained by using the vibro-rotational transitions of molecular hydrogen, since H_2 is a very abundant molecule although very rarely seen in quasar absorber. Very few studies used other molecules since they are difficult to detect and measure accurately at large redshifts. In general these methods provide less stringent bounds for high redshifts. This thesis will concentrate on the single H_2 system observed towards QSO 0347-383 to trace the proton-to-electron mass ration μ at high redshift ($z_{\rm abs} = 3.025$). This work reaches a robust estimation of the achievable accuracy with current data by comparing independent observation runs and the latest available data.

2 Background

2.1 Cosmology of varying fundamental constants

The development of physics relied considerably on the Copernican principle, which states that the Earth is not in a central, specially favored position in the universe and that the laws of physics do not differ from one point in spacetime to another. In cosmology, if one assumes the Copernican principle and observes that the universe appears isotropic from our vantage-point on Earth, then one can prove that the Universe is generally homogeneous (at any given time) and is also isotropic about any given point. These two conditions comprise the cosmological principle. It is however natural to question this assumption. It is difficult to imagine a change of the form of physical laws but a smooth change in the physical constants is much easier to conceive. Comparing and reproducing experiments is also a root of the scientific approach which makes sense only if the laws of nature do not depend on time and space. This hypothesis of constancy of the constants plays an important role in particular in astronomy and cosmology where the redshift measures the look-back-time. Ignoring the possibility of varying fundamental physical constants could lead to a distorted view of our universe and if such a variation is established corrections would have to be applied. It is thus of great importance to investigate this possibility especially as the measurements become more and more precise.

Evidently, the constants have not undergone huge variations on Solar system scales and geological time scales and one is looking for tiny effects. The question of the numerical values of fundamental physical constants is central to physics and one can hope to explain them dynamically as predicted by some high-energy theories. Testing the constancy of the constants is part of the tests of general relativity.

This speculative theory which embeds varying constants is analogous to the transition from the Newtonian description of mechanics in which space and time were just a static background in which matter was evolving to the relativistic description where spacetime becomes a dynamical quantity determined by the Einstein equations (Damour 2001).

There are several reasons why the possibility of varying constants should be taken seriously. First, we know that the best candidates for unification of the forces of nature in a quantum gravitational environment only seem to exist in finite form if there are many more dimensions of space than the three that we are familiar with. This means that the true constants of nature are defined in higher dimensions and the three-dimensional projections we observe are no longer fundamental and do not need to be constant. Any slow change in the scale of the extra dimensions would be revealed by measurable changes in our three-dimensional 'constants'. Second, we appreciate that some apparent constant might be determined partially or completely by spontaneous symmetry-breaking processes in the very early universe.

This introduces an irreducibly random element into the values of those constants. They may be different in different parts of the universe and hence at different directions or different redshifts. The most dramatic manifestation of this process is provided by the chaotic and eternal inflationary universe scenarios where both the number and the strength of forces in the universe at low energy can fall out differently in different regions. Third, any outcome of a theory of quantum gravity will be intrinsically probabilistic. It is often imagined that the probability distributions for observables will be very sharply peaked but this may not be the case for all possibilities. Thus, the value of the gravitation 'constant', G, or its time derivative, $\dot{G}$, might be predicted to be spatial random variables. Fourth, a non-uniqueness of the vacuum state for the universe would allow other numerical combinations of the constants to have occurred in different places. String theory indicates that there is a huge 'landscape' ($> 10^{500}$) of possible vacuum states that the universe can find itself residing in as it expand and cools (Barrow 2005). Each will have different constants and associated forces and symmetries. It is sobering to remember that at present we have no idea why any of the natural constants take the numerical values they do and we have never successfully predicted the value of any dimensionless constant in advance of its measurement.

A fist step is to evaluate which physical constants are to be considered in general. Lévy-Leblond (1977) defined three classes of fundamental constants, since not all constants of physics play the same role:

- The class A of the constants characteristic of particular objects,
- The class B of the constants characteristic of a class of physical phenomena,
- The class C being the class of universal constants.

This definition of a fundamental constant, however, can cause the change of status of constants, as can be exemplified by the constant c, the speed of light. Initially being a type A constant (describing a property of light), then becoming a type B constant when it was realized that it was related to the electro-magnetic phenomena and it ended as type C constant (it is part of many laws of physics from electromagnetism to relativity). It has even become a much more fundamental constant since it has been chosen as the new definition of the meter (see Petley 1983).

A more conservative definition of a fundamental constant would thus be to state that it is any parameter that can not be calculated with our present knowledge of physics, e.g. a free parameter of our theory at hand. Each free parameter of any theory is in fact a challenge for future theories to explain the value (Uzan 2003).

The set of constants which are conventionally considered as fundamental consists of the electron charge e, the electron mass m_e, the proton mass m_p, the reduced Planck constant $\hbar$, the velocity of light in vacuum c, the Avogadro constant N_A, the Boltzmann constant k_B, the

Newton constant G, the permeability and permittivity of space, ε_0 and μ_0. The latter has a fixed value in the SI system of unit ($\mu_0 = 4\pi \times 10^{-7}\,\text{H}\,\text{m}^{-1}$) which is implicit in the definition of the Ampere; ε_0 is then fixed by the relation $\varepsilon_0 \mu_0 = c^{-2}$. To compare with, the minimal standard model of particle physics plus gravitation that describes the four known interactions depends on 20 free parameters (Cahn 1996; Hogan 2000): the Yukawa coefficients determining the masses of the six quark (u, d, c, s, t, b) and three lepton (e, μ, τ) flavors, the Higgs mass and vacuum expectation value, three angles and a phase of the Cabibbo-Kobayashi-Maskawa matrix, a phase for the QCD vacuum and three coupling constants g_s, g_w, g_1 for the gauge group $SU(3) \times SU(2) \times U(1)$ respectively. Below the Z mass, g_1 and g_w combine to form the electro-magnetic coupling constant.

The final number of free parameters indeed depends on the physical model at hand (see, e.g., Weinberg 1983). The introduction of constants in physical law is closely related to the existence of systems of units. Newton's law states that the gravitational force between two masses is proportional to each mass and inversely proportional to their separation. To transform the proportionality to an equality one requires the use of a quantity with dimension of $\text{m}^3\text{kg}^{-1}\text{s}^{-2}$ independent of the separation between the two bodies, of their mass, of their composition (*equivalence principle*) and on the position (*local position invariance*). With another system of units this constant could have simply been anything.

The determination of the laboratory value of constants relies mainly on the measurements of lengths, frequencies, times,... (see Flowers and Petley 2001). Hence, any question on the variation of constants is linked to the definition of the system of units and to the theory of measurement. The choice of a base units affects the possible time variation of constants.

The behavior of atomic matter is mainly determined by the value of the electron mass and of the fine structure constant. The Rydberg energy sets the (non-relativistic) atomic levels, the hyperfine structure involves higher powers of the fine structure constant, and molecular modes (including vibrational, rotational modes) depend on the mass ratio m_p/m_e. As a consequence, if the fine structure constant is spacetime dependent, the comparison between several devices such as clocks and rulers will also be spacetime dependent. This dependence will also differ from one clock to another so that metrology becomes both device and spacetime dependent.

Besides this first metrologic problem, the choice of units has implications on the permissible variations of certain dimensionful constant. Petley (1983) discusses the implication of the definition of the meter for example. The original definition of the meter via a prototype platinum-iridium bar depends on the interatomic spacing in the material used in the construction of the bar. Atkinson (1968) argued that, at first order, it mainly depends on the Bohr radius of the atom so that this definition of the meter fixes the combination (2.12) as constant. Another definition was based on the wavelength of the orange radiation from krypton-86 atoms. It is likely that this wavelength depends on the Rydberg constant and on the reduced mass of the atom so that it ensures that $m_e c^2 \alpha_{\text{EM}}^2 / 2\hbar$ is constant. The more recent definition of the meter as the length of the path traveled by light in vacuum during a time of $1/299,792,458$ of a second imposes the constancy of the speed of light[1] c. Identically, the definitions of the second as the duration of 9,192,631,770 periods of the transition between two hyperfine levels of the ground state of cesium-133 or of the kilogram via an international prototype respectively impose that $m_e^2 c^2 \alpha_{\text{EM}}^4 / \hbar$ and m_p are fixed.

Since the definition of a system of units and the value of the fundamental constants (and thus the status of their constancy) are entangled, and since the measurement of any dimensionful quantity is in fact the measurements of a ratio to standards chosen as units, *it only makes sense to consider the variation of dimensionless ratios*.

The required approach is to focus on the variation of dimensionless ratios which, for instance, characterize the relative magnitude of two forces, and are independent of the choice of the system of units and of the choice of standard rulers or clocks.

Notations: In this work, SI units and the following values of the fundamental constants today[2] are used:

$$c = 299,792,458 \, \text{ms}^{-1} \tag{2.1}$$
$$\hbar = 1.054571596(82) \times 10^{-34} \, \text{Js} \tag{2.2}$$
$$G = 6.673(10) \times 10^{-11} \, \text{m}^3 \text{kg}^{-1} \text{s}^{-2} \tag{2.3}$$
$$m_e = 9.10938188(72) \times 10^{-31} \, \text{kg} \tag{2.4}$$
$$m_p = 1.67262158(13) \times 10^{-27} \, \text{kg} \tag{2.5}$$
$$m_n = 1.67492716(13) \times 10^{-27} \, \text{kg} \tag{2.6}$$
$$e = 1.602176462(63) \times 10^{-29} \, \text{C} \tag{2.7}$$

for the velocity of light, the reduced Planck constant, the Newton constant, the masses of the electron, proton and neutron, and the charge of the electron.

Defined as well are

$$q^2 \equiv \frac{e^2}{4\pi\varepsilon_0}, \tag{2.8}$$

and the following dimensionless ratios

$$\alpha_{\text{EM}} \equiv \frac{q^2}{\hbar c} \sim 1/137.03599976(50) \tag{2.9}$$
$$\mu \equiv \frac{m_p}{m_e} \sim 1836.15267247(80). \tag{2.10}$$

$$\tag{2.11}$$

---

[1] Note that the velocity of light is not assigned a fixed value *directly*, but rather the value is fixed as a consequence of the definition of the meter.

[2] see http://physics.nist.gov/cuu/Constants/ for an up to date list of the recommended values of the constants of nature.

The notations

$$a_0 = \frac{\hbar}{m_e c \alpha_{\text{EM}}} = 0.5291771 \text{ Å} \quad (2.12)$$

$$-E_I = \frac{1}{2} m_e c^2 \alpha_{\text{EM}}^2 = 13.60580 \text{ eV} \quad (2.13)$$

$$R_\infty = -\frac{E_I}{hc} = 1.0973731568549(83) \times 10^7 \text{ m}^{-1} \quad (2.14)$$

respectively for the Bohr radius, the hydrogen ionization energy and the Rydberg constant are introduced.

Note, in some works μ is referred to as electron-to-proton mass ratio m_e/m_p, which has the effect of a change in sign for $\Delta\mu/\mu$. The cited values in this thesis are converted accordingly.

2.1.1 Accessible constants

A prominent fundamental constant that meets the above mentioned requirements is the proto-to-electron mass ratio $\mu = m_p/m_e$ (see Eq. 2.10). The time variation of μ is given by:

$$\frac{\dot{\mu}}{\mu} = \frac{\dot{m}_p}{m_p} - \frac{\dot{m}_e}{m_e}. \quad (2.15)$$

Though the proton mass m_p depends not only on the quantum chromodynamics (QCD) scale Λ_{QCD} but also on the masses of the up quark and the down quark, m_p is usually considered to be proportional to Λ_{QCD} since these quark masses are much smaller than Λ_{QCD}.

The fine-structure constant α (see Eq. 2.9) is not taken into account here but has proven to be another suitable fundamental physical constant. For an atom/ion, the relativistic corrections to the energy levels of an electron are proportional to α^2, although the magnitude of the change depends on the transition under consideration. The tests for variation in μ and α run completely independent from each other but most theories suggest a certain correlation between the two. The fine-structure constant is hence mentioned since early observations gave rise to this rich field of varying constants and the efforts in constraining both α and μ stand to benefit from each other.

Theoretically there is a wide range of possible connections between the fine-structure constant α and the proton-to-electron mass ratio μ. In numerous considered models variations in α lead to variations in the electron mass (via the electron self-energy) and in the proton mass (via the electrostatic energy contained inside a proton). These are model dependent and in general quite complex to work out (see, e.g., Dine et al. 2003).

The first observational indications of potential variation stimulated numerous theoretical works to explain the new findings. An often-cited paper[3] on α is the one by Webb et al. (2001), which reports:

$$\frac{\Delta\alpha}{\alpha} = (-0.72 \pm 0.18) \times 10^{-5}, \quad (2.16)$$

---
[3]More than 430 citations to this day.

Or later, a paper on α by Murphy et al. (2003):

$$\frac{\Delta\alpha}{\alpha} = (-0.543 \pm 0.116) \times 10^{-5}, \tag{2.17}$$

though similar observations of other groups did not necessarily reproduce that result. Observations of Reinhold et al. (2006) suggested a fractional change in the proton-to-electron mass ratio $\mu = m_\mathrm{p}/m_\mathrm{e}$:

$$\frac{\Delta\mu}{\mu} = (2.4 \pm 0.6) \times 10^{-5}, \tag{2.18}$$

for a weighted fit to observations at a redshift of $z \sim 3$, which implies that the proton-to-electron mass ratio has decreased over the last 12 Gyr. These two extremes among the different findings were taken as boundary conditions for a large range of theoretical works, since from the theoretical point of view, it is natural to allow time and space dependence of fundamental constants. In fact, superstring theory, which is expected to unify all fundamental interactions, predicts the existence of a scalar partner ϕ (called dilaton) of the tensor graviton, whose expectation value determines the string coupling constant $g_s = e^{\phi/2}$ (Witten 1984). The couplings of the dilaton to matter induces the violation of the equivalence principle and hence generates deviations from general relativity. With the above mentioned observational results, the hints of the time variation of fundamental constants were considered to be found (see, e.g., Calmet and Fritzsch 2006; Chiba et al. 2007).

The following section will specify the modus operandi and the necessary requirements to measure the proton-to-electron mass ratio on cosmological scales.

2.2 Observables

2.2.1 Proton-to-electron mass ratio μ

As first pointed out by Thompson (1975) molecular absorption lines can provide a test of the variation of μ. The energy difference between two adjacent rotational levels in a diatomic molecule is proportional to Mr^{-2}, r being the bond length and M the reduced mass, and that the vibrational transition of the same molecule has, in first approximation, a $\sqrt{M}$ dependence. For molecular hydrogen $M = m_\text{p}/2$ so that comparison of an observed vibro-rotational spectrum with its present analog will thus give information on the variation of m_p and m_n. Comparing pure rotational transitions with electronic transitions gives a measurement of μ.

Following Thompson (1975), the frequency of vibration-rotation transitions is, in the Born-Oppenheimer approximation, of the form

$$\nu \sim E_I \left(c_\text{elec} + c_\text{vib}/\sqrt{\mu} + c_\text{rot}/\mu \right) \tag{2.19}$$

where c_elec, c_vib and c_rot are some numerical coefficients. Comparing the ratio of wavelengths of various electronic-vibration-rotational lines in quasar spectrum and in the laboratory allow to trace the variation of μ since, at lowest order, Eq. (2.19) implies

$$\frac{\Delta E_{ij}(z)}{\Delta E_{ij}(0)} = 1 + K_{ij} \frac{\Delta \mu}{\mu} + \mathcal{O}\left(\frac{\Delta \mu^2}{\mu^2}\right), \tag{2.20}$$

where the coefficients K_{ij} determine the sensitivity of the transition energies to a change in μ. An important point is that the values of K_{ij} differ for different lines. Thus, if the reduced mass of a molecule at the epoch z differs from the present value, then the observered wavelength and the corresponding sensitivity coefficient K_{ij} of that transition must be linearly correlated. This implicit correlation underlies the method. Section 2.2.4 describes how these coefficients can be computed.

2.2.2 Intergalactic H_2

Molecular hydrogen H_2 is the most abundant molecule in the universe and plays a fundamental role in many astrophysical contexts. It is found in all regions where the shielding of the ultraviolet photons, responsible for the photo-dissociation of H_2, is sufficiently large. Except in the early universe, most H_2 is thought to be produced via surface reactions on interstellar dust grains, since gas-phase reactions are too slow in general (see, e.g., Habart et al. 2004).

The H_2 formation mechanism is not yet fully understood. Direct observations of H_2 are difficult since electronic transitions occur only in the ultraviolet to which Earth's atmosphere is opaque. UV satellites are only suited for bright nearby objects and could not provide the necessary resolution. Only at great distances and hence with a large redshift these spectra are shifted into the visual band and can then be observed with ground-based telescopes. Another problem is the narrow range of conditions under which H_2 forms. The required dust grains that allow for the forming of molecular hydrogen can easily obscure the molecular hydrogen as well.

Hydrogen makes up about 80% of the known matter in the universe and most of it is contained in either atomic or molecular hydrogen in the gaseous phase (Combes and Pineau Des Forets 2000). It took until 1970 for the first detection of molecular hydrogen in space; the observation was made possible through the use of a rocket borne spectrometer observing from high altitudes, therewith evading atmospheric absorption of the vacuum ultraviolet radiation. Lyman bands in the wavelength range between 1000 and 1100 Å were identified in the absorption spectrum of a diffuse interstellar cloud in the optical path towards ξ Persei (Carruthers 1970). Further satellite based observations also revealed absorption of Werner bands and UV emission of Lyman and Werner bands including their continua (Spitzer et al. 1974). The Copernicus satellite telescope greatly improved the possibilities for recording UV-spectra of molecular hydrogen (see, e.g., Morton and Dinerstein 1976).

The International Ultraviolet Explorer (IUE), launched January 1978 and in service until September 1996, covered ultraviolet wavelengths from 1200 to 3350 Å with two on-board spectrographs. It detected for example vibrationally excited molecular hydrogen in the upper atmosphere of Jupiter (Cravens 1987).

The new Far Ultraviolet Spectroscopic Explorer (FUSE), in flight between June 1999 and october 2007, is an ideally suited spectroscopic measurement device to probe hydrogen in space. It covers the wavelength range 905-1187 Å, the range of the strong Lyman and Werner absorption bands, with high resolution and it is now used routinely for H_2 observations (Moos et al. 2000).

For this thesis, publicly available FUSE data was widely used to verify line lists of vibro-rotational transitions and to test the written graphical data examination (GRADE[4]) tool against. However, its data unfortunately cannot be used for local space based measurements of μ since the FUSE satellite bears no on board calibration set up. Instead the obtained spectra are calibrated via the observed H_2 absorption features.

The abundance of molecular hydrogen in space is usually expressed as the fraction $f(H_2) \equiv 2N(H_2)/[2N(H_2) + N(H\,I)]$.

Savage et al. (1977) found the correlation for H_2 in our galaxy:

$$f(H_2) \geq 10^{-2} \text{ for } N(H\,I) > 4 \times 10^{20}\,\text{cm}^{-2}. \tag{2.21}$$

A threshold of 5 Å for the equivalent width in the search for DLA systems, as mentioned above, corresponds to $N(H\,I) \geq 2 \times 10^{20}\,\text{cm}^{-2}$ and thus a fraction of $f(H_2) > 10^{-2}\,\text{cm}^{-2}$ would be expected for most of the DLA systems. Albeit the observed $f(H_2)$ in distant DLA is much lower than that. The low H_2 content in DLA in contrast to our galaxy is likely due to their comparably low dust contents. The fraction of molecular Hydrogen can be described as the ratio of its formation on dust grains and its photodissociation by UV-photons via $f(H_2) = 2Rn/I$

---

[4] see Appendix for a short summary of programs written in the course of this thesis.

Table 2.1: List of damped Lyman-α systems with H_2 absorption observations

Quasar source	redshift z_{abs}
QSO 0515-441	1.15
QSO 1331+170	1.78
QSO 0551-336	1.96
QSO 0013-004	1.97
QSO 1444+014	2.09
QSO 1232+082	2.34
QSO 2343+125	2.43
QSO 0405-443	2.59
QSO 0528-250	2.81
QSO 0347-383	3.02
QSO 0000-263	3.39
QSO 1443-272	4.22

where I is proportional to the intensity of UV radiation and R is proportional to the dust-to-gas ratio κ. A low $f(H_2)$ in DLAs could then be attributed to a low dust content, and thus a low κ. Such a correlation was indeed found by Petitjean et al. (2002).

Long before the actual observation of molecular hydrogen, Herzberg had discussed the possibility of detecting H_2 in outer space through the quadrupole spectrum, even before these very weak features were observed in the laboratory (Herzberg 1949). With the further development of infrared sensitive CCD cameras, the 2 μm infrared emissions, coinciding with an atmospheric transmission window, could be mapped in 2D-imaging pictures of distributions of hot molecular hydrogen in space (Field et al. 1994).

Levshakov and Varshalovich (1985) tentatively assigned some features in spectra obtained by Morton et al. (1980) from PKS 0528-250 (one of the few systems up to date used for determination of μ). Similar spectra of this system were collected by Foltz et al. (1988) and this data formed the basis for a constraint on a possible variation of μ put forward by Varshalovich and Levshakov (1993).

Further reports on observation of molecular hydrogen absorption lines at high redshift are given by Ge and Bechtold (1997) at $z = 1.97$ towards QSO 0013-004, by Reimers et al. (2003) at $z = 1.15$ towards QSO 0515-441, and by Cui et al. (2005) at $z = 1.78$ towards QSO 1331+170. Observations by the VLT/UVES instrument led to H_2 detection towards QSO 0347-383, QSO 1232+082 (Ivanchik et al. 2002; Levshakov et al. 2002) and towards QSO 0551-336 (Ledoux et al. 2002).

Additional observations of H_2 are reported towards Q 0000-263 in Levshakov et al. (2000). Ledoux et al. (2003) and Srianand et al. (2005) performed surveys on Damped Lyman-a (DLA) systems at redshifts $z > 1.8$, in which some new quasars with H_2 absorption were detected.

From their study and from past searches they conclude that molecular hydrogen is detected in 13 − 20% of the systems.

More recently Petitjean et al. (2006) observed the systems QSO 2343+125 and QSO 2348-011, while Ledoux et al. (2006) observed H_2 lines in a source at the highest redshift until now ($z = 4.22$).

The observations of 2006 and the continued survey for DLAs demonstrate that the amount of known H_2 absorbing clouds at high redshift is rapidly expanding; it is therefore likely that additional high resolution data to extract information on μ variation will become available in the near future. Noterdaeme et al. (2008) at all conclude from the comparison between H_2-bearing systems and the overall UVES sample, that a significant increase of the molecular fraction in DLAs could take place at redshifts $z_{\text{abs}} \geq 1.8$. The known DLA systems with H_2 absorption are listed in Table 2.1.

2.2.3 Laboratory wavelengths

For some time the only available precise data on oscillator frequencies and for this study more important rest frame wavelengths for molecular hydrogen were those computed by Abgrall et al. (1993a). The claimed accuracy lay at about 1m Å corresponding to $\sim$ 4 mÅ for the observer's frame in this case. This is on the order of the by now reached accuracy in line fits and improvements in the determination of the restframe wavelength was mandatory.

Philip et al. (2004) eventually conducted high-resolution laser-spectroscopy to gain precise transition frequencies in the Lyman and Werner bands via direct measurements. A strong test on the accuracy of transition frequencies is to compare the differences between the rotational branches $P(J+2)$ and $R(J)$. They should match the calculated ground state rotational splittings, which are accurately known (see Jennings et al. 1984). The achieved accuracy is stated as $<$ 0.01 mÅ. In the framework of this analysis the rest frame wavelength can thus be assumed to be exact. Due to experimental restrictions on the UV laser range the transition frequencies were obtained only for a subset of the lines detected in the spectrum of QSO 0347-383. As can be seen in Figure 2.1 the new data has a throughout positive varying offset, which strongly influences $\Delta\mu/\mu$ analysis.

More recent measurements (Hollenstein et al. 2006; Ivanov et al. 2008; Salumbides et al. 2008; Bailly et al. 2010) completed the data on the Lyman and Werner band frequencies. The new data tables include all observed and selected H_2 lines. At the time, Ubachs made the data available prior to publication via private communication so they already could be used in the present study right from the start. Figure 2.1 illustrates the increasing uncertainty in calculations for shorter wavelengths or rather higher vibrational levels. The deviation of calculated values and laboratory measurements is evident and caused early inconsistent findings (see Eq. 2.34 and 2.35 in section 2.2.5). The stated errorbars of $<$ 0.01 mÅ for the new data are below the size of the data points in the plot. The refinements after 2004 are notedly below that even.

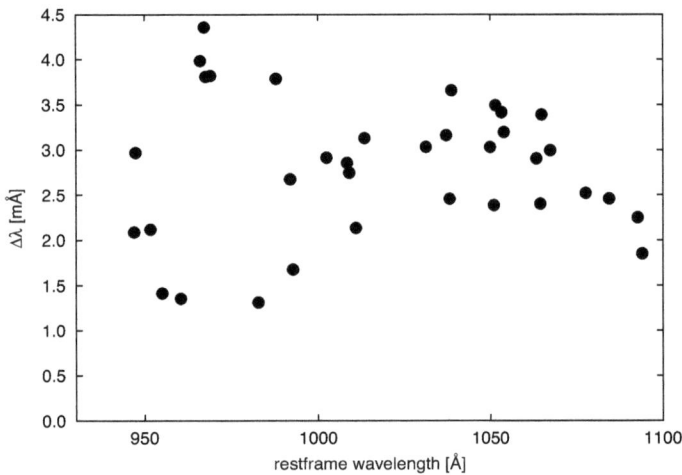

Figure 2.1: Changes in mÅ in rest frame wavelengths $\Delta\lambda = \lambda_P - \lambda_A$ between data by Philip et al. (2004) and Abgrall et al. (1993a), "P" and "A" respectively.

2.2.4 Sensitivity coefficients K_i

The coefficients have been firstly calculated by Varshalovich and Levshakov (1993) from Dunham's spectroscopic constants for the H_2 molecule using theoretical ideas about the dependence. As mentioned in section 2.2.1, electronic, vibrational, and rotational excitations of a diatomic molecule depend differently on its reduced mass M and hence on the proton-to-electron mass ratio μ for molecular hydrogen. To a first approximation, these energies are proportional to μ^0, $\mu^{-\frac{1}{2}}$, μ^{-1}, respectively. Hence each transition has an individual sensitivity to a possible change in that reduced mass. This can be expressed by a sensitivity coefficient.

$$K_i = \frac{d\ln\lambda_i}{d\ln\mu} = \frac{\mu}{\lambda_i}\frac{d\lambda_i}{d\mu}. \quad (2.22)$$

A first estimation of these coefficients can be obtained by comparing transitions of molecular hydrogen with deuterium or tritium. Transitions with equal rotational and vibrational quantum numbers have different energies for H_2 and deuterium, or tritium. Recently new experimental data on molecular hydrogen and deuterium level energies were obtained by sophisticated laboratory measurements by Hollenstein et al. (2006).

Since H_2 and D_2 classically only differ in mass, K_i was initially computed for this work via Equation 2.22 for each transition using the available line data of H_2 and deuterium (`KiComp`). Of course this is a simplified approach. Figure 2.2 shows a comparison between the coefficients calculated via the mentioned method in thesis and up to date values. As can be seen they are in rather good agreement for the longer wavelengths or lower vibrational levels, respectively.

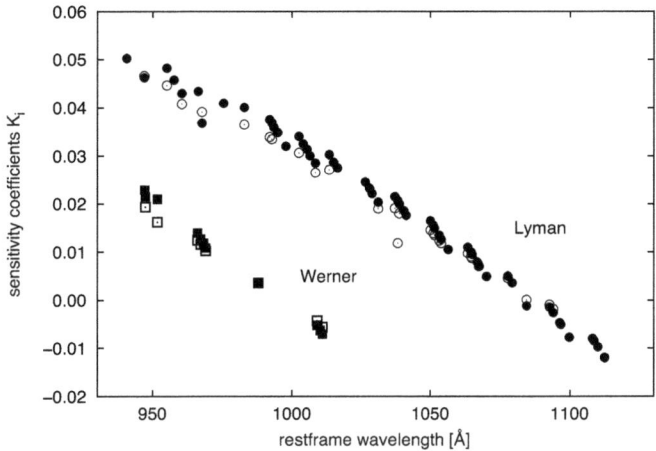

Figure 2.2: Sensitivity coefficients of observed lines in the Lyman (*circles*) and Werner (*squares*) band as calculated via deuterium (*open symbols*) and by Ubachs et al. (2007) (*solid*).

Reinhold et al. (2006) refined the calculations of K_i. In first order they can be expressed by the Dunham coefficients Y_{kl} of the ground and excited states. With $\mu_n = \frac{m_e\mu}{2}$, Equation 2.22 leads to:

$$K_i = -\frac{\mu_n}{\lambda_i}\frac{d\lambda_i}{d\mu_n} = \frac{1}{E_e - E_g}\left(-\frac{\mu_n dE_e}{d\mu_n} + \frac{\mu_n dE_g}{d\mu_n}\right). \tag{2.23}$$

The computations for the energies of the excited and ground state, E_e and E_g, respectively are the same as for the energy levels of H_2. Starting with the Born-Oppenheimer approximation (BOA) based on the semi empirical approach the energy levels can be expressed by the Dunham formula (see Dunham 1932):

$$E(v, J) = \sum_{k,l} Y_{kl}\left(v + \frac{1}{2}\right)^k [J(J+1) - \Lambda^2]^l; \qquad \Lambda^2 = \begin{cases} 0 \text{ for Lyman} \\ 1 \text{ for Werner} \end{cases} \tag{2.24}$$

However, the Dunham coefficients Y_{kl} cannot be calculated directly from the level energies due to strong mutual interaction between the excited states as well as avoided rotational transitions between nearby vibrational levels. For the first time the more complex non-BOA effects are taken into account in Reinhold et al. (2006).

Ubachs et al. (2007) further improved the accuracy of sensitivity coefficients with laboratory measurements of the level energies of molecular hydrogen via XUV-laser experiments allowed for a reliable enhancement of the BOA approximation. The Dunham coefficients Y_{kl} for the lower states were fitted via experimental data. In general the sensitivity coefficients are largest at the shortest wavelengths (see Figure 2.2), for both the Lyman and Werner systems. This is explained from the high vibrational quantum numbers associated with those lines. Further it can be noted that for each band system at the long wavelength side the K_i values become negative. This is due to the larger zero-point vibrational energy in the ground state than in the excited states.

Assessing the accuracy of these sensitivities proves to be very difficult. Ubachs et al. (2007) estimate the overall uncertainty to be within 5×10^{-4}, which corresponds to 1% of the full range of K_i values (between -0.01 and 0.05). A more profound test can be accomplished though. Almost simultaneous to the efforts by Reinhold et al. (2006), *ab initio* calculations of the sensitivity coefficients were carried out by Meshkov et al. (2006). The differences between the coefficients

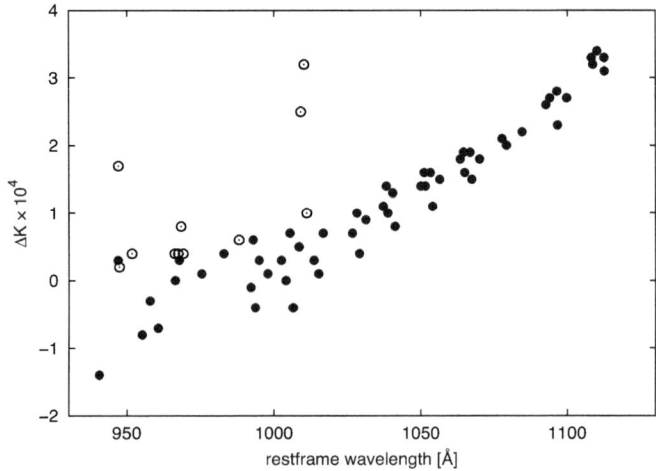

Figure 2.3: Differences $\Delta K = K_{SE} - K_{AI}$ between derived K coefficients of the semi-empirical approach (Ubachs et al. 2007) and those of the ab initio calculation in Meshkov et al. (2006) for the Lyman (*solid*) and Werner (*open*) band.

from the semi-empirical analysis (K_{SE}), and the completely independent values (K_{AI}) from ab initio analysis are plotted is Figure 2.3. All deviations ΔK lie within margins of -2×10^{-4} and $+4 \times 10^{-4}$, corresponding to less than 1% of the range that the K_i values exhibit. In view of the entirely independent approaches to the problem this comparison produces some confidence in the correctness of the derived values. The plot however indicates systematic deviations for the Werner band (*open circles*) and a general increase in disagreement for the lower vibrational levels. At the current situation the estimated errors in the sensitivity coefficients have no impact on the final result as shown in section 5.3.

2.2.5 Status quo for $\Delta\mu/\mu$

Varshalovich and Levshakov (1993) used the observations of a damped Lyman-α system associated with the quasar PKS 0528-250 of redshift $z = 2.811$ and deduced that

$$|\Delta\mu/\mu| < 4 \times 10^{-3}. \tag{2.25}$$

A similar analysis was first tried by Foltz et al. (1988) but their work did not take into account the wavelength-to-mass sensitivity and their result hence seems not very reliable. Nevertheless, they concluded that for $z = 2.811$:

$$|\Delta\mu/\mu| < 2 \times 10^{-4}. \tag{2.26}$$

Cowie and Songaila (1995) observed the same quasar and deduced that

$$\Delta\mu/\mu = (-0.75 \pm 6.25) \times 10^{-4}, \tag{2.27}$$

at 95% C.L. from the data on 19 absorption lines.
Varshalovich and Potekhin (1995) calculated the coefficient K_{ij} to a higher precision and deduced that

$$|\Delta\mu/\mu| < 2 \times 10^{-4}. \tag{2.28}$$

Thereinafter, Varshalovich et al. (1996) used 59 transitions for H_2 rotational levels in PKS 0528-250 and got

$$\Delta\mu/\mu = (10 \pm 12) \times 10^{-5}, \tag{2.29}$$

at 2σ level.
These results were confirmed by Potekhin et al. (1998) using 83 absorption lines to get

$$\Delta\mu/\mu = (7.5 \pm 9.5) \times 10^{-5}, \tag{2.30}$$

at a 2σ level.
Later, Ivanchik et al. (2001) measured, with the VLT, the vibro-rotational lines of molecular hydrogen for two quasars with damped Lyman-α systems respectively at $z = 2.3377$ and $z = 3.0249$ and also argued for the detection of a time variation of μ. Their most conservative result is (the observational data were compared to two experimental data sets)

$$\Delta\mu/\mu = (5.7 \pm 3.8) \times 10^{-5}, \tag{2.31}$$

at 1.5σ and the authors cautiously point out that additional measurements are necessary to ascertain this conclusion. The result is also dependent on the laboratory dataset of transition frequencies used for the comparison since it gave $\Delta\mu/\mu = (12.2 \pm 7.3) \times 10^{-5}$ with another dataset.
As in the case of Webb et al. (2001, 1999), indicating a detected variation in $\alpha_{\rm EM}$, this measurement is very important in the sense that it is a non-zero detection that will have to be compared

with other bounds. The measurements by Ivanchik et al. (2001) is indeed much larger than one would expect from the electromagnetic contributions. As seen in section 2.1 for any unified theory the changes in the masses are expected to be larger than the change in α_{EM}. Typically, we expect $\Delta\mu/\mu \sim \Delta\Lambda_{\text{QCD}}/\Lambda_{\text{QCD}} - \Delta v/v \sim (30-40)\Delta\alpha_{\text{EM}}/\alpha_{\text{EM}}$, so that it seems that the detection by Webb et al. (2001) is too large by an order of magnitude to be compatible with it (Uzan 2003).

Levshakov et al. (2002) identified more than 80 H_2 molecular lines in a damped Lyα (DLA) system at $z_{\text{abs}} = 3.025$ toward QSO 0347-383. Due to HI Lyα forest contamination several were considered unsuitable for further analysis and a subset of 15 lines were chosen to set an upper limit on possible changes of μ:

$$|\Delta\mu/\mu| < 5.7 \times 10^{-5}. \tag{2.32}$$

Ivanchik et al. (2003) find for QSO 0347-383:

$$\Delta\mu/\mu = (5.02 \pm 1.82) \times 10^{-5}. \tag{2.33}$$

In general the given errors represent the statistical errors alone. Which becomes evident in the follow up investigations of the same system:

Based on the wavelengths given by Abgrall et al. (1993a,b) Ivanchik et al. (2005) find:

$$\Delta\mu/\mu = (3.05 \pm 0.75) \times 10^{-5}, \tag{2.34}$$

or, using new laboratory measurements by Philip et al. (2004) for wavelengths data:

$$\Delta\mu/\mu = (1.65 \pm 0.74) \times 10^{-5}, \tag{2.35}$$

and eventually the result by Reinhold et al. (2006):

$$\Delta\mu/\mu = (2.4 \pm 0.6) \times 10^{-5}. \tag{2.36}$$

For the sake of completeness it should be noted, that Pagel (1983) used another method to constrain μ based on the measurement of the mass shift in the spectral lines of heavy elements. In that case the mass of the nucleus can be considered as infinite contrary to the case of hydrogen. A variation of μ will thus influence the redshift determined from hydrogen. He compared the redshifts obtained from spectrum of hydrogen atom and metal lines for quasars of redshift ranging from 2.1 to 2.7. Since

$$\Delta z \equiv z_{\text{H}} - z_{\text{metal}} = (1+z)\frac{\Delta\mu}{1-\mu_0}, \tag{2.37}$$

he obtained that

$$|\Delta\mu/\mu| < 4 \times 10^{-1}, \tag{2.38}$$

at $3\,\sigma$ level. This result is unfortunately not conclusive because usually heavy elements and hydrogen belong to different interstellar clouds with different radial velocity.

Apparently the laboratory measurements of μ itself were refined over the same period as Figure 2.4 illustrates.

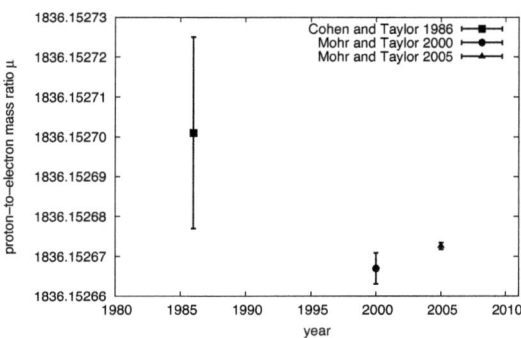

Figure 2.4: Measurements of the proton-to-electron mass ratio, representing the values for μ listed by the National Institute of Standards and Technology (NIST).

3 Analysis I

3.1 Data

3.1.1 QSO 0347-383

The source of the analysed spectrum is a bright quasi-stellar radio object (QSO) with a visual magnitude of $V = 17.3$ mag at a redshift of $z = 3.23$ (Maoz et al. 1993), which shows a Damped Lyman α system (DLA) at $z_{abs} = 3.0245$. The hydrogen column density is N(H I)= 5×10^{20} cm^{-2} with a rich absorption-line spectrum (Levshakov et al. 2002). The DLA exhibits a multicomponent velocity structure. There are at least two gas components: warm gas seen in lines of neutral atoms, H and low ions, and hot gas where the resonance doublets of C IV and Si IV are formed. In the cooler component molecular hydrogen was first detected by Levshakov et al. (2002) who identified 88 H$_2$ lines. First High-resolution spectra of the quasar QSO 0347-383 were obtained with the Ultraviolet-Visual Echelle Spectrograph (UVES) during commissioning at the Very Large Telescope (VLT) 8.2m ESO telescope by D'Odorico et al. (2001). QSO 0347-383 is the identifier of the "Fundamental-Katalog 4.0" calibrated to 1950 but still being widely used. The precise position dated to 2000 as stated in the fifth fundamental catalogue is α = 03h 49m 43.68s, δ = -38°10' 31.3".

QSO 0347-383 itself was discovered by Osmer and Smith (1980). For the present analysis two independent data sets are taken into account. This first one was already described by Ivanchik et al. (2005) and Wendt and Reimers (2008). The Quasar absorption line spectra were obtained with the UVES spectrograph at the Very Large Telescope (VLT) of the European Southern Observatory (ESO) in Paranal, Chile. The slit was 0.8 arcsec wide resulting in a spectral resolution of $R \sim 53.000$ over the wavelength range 3300 Å– 4500 Å.

The average seeing during observation was about 1.2 arcsec. Before and after the exposures for each night, Thorium-Argon calibration data were taken. An overall of 9 spectra were recorded with an exposure time of 4500 seconds each between January 8th and January 10th 2002 for the ESO program 68.A-0106(A). All spectra were taken with grating with a central wavelength of 4303 Å and the blue "Pavarotti"-CCD with 2 × 2 binning. Later on the data were reduced manually by Mirka Dessauges-Zavadsky from Geneva Observatory in Jan 2004 to achieve maximum accuracy. The ESO Ambient Conditions Database[1] includes measurements of the environmental parameters at the Paranal ESO observatory and shows no significant

---
[1] http://archive.eso.org/eso/ambient-database.html

changes in temperatures during or in between the exposures that could lead to shifts between the separate observations. All works on QSO 0347-383 are based on the same above mentioned

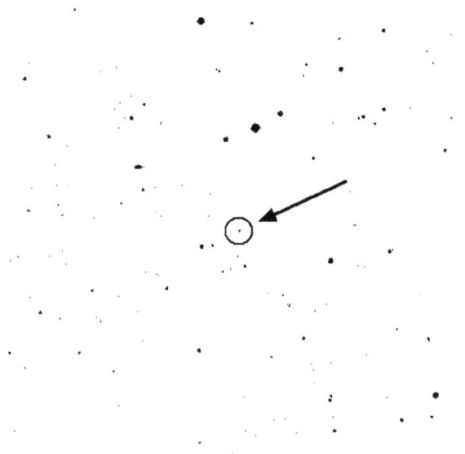

Figure 3.1: Colour inverted and contrast enhanced photograph taken in the Blue-Band (J) covering a 14' × 14' area. QSO 0347-383 is marked by a circle and arrow. Original image from Space Telescope Science Institute (STScI).

UVES VLT observations[2] in January 2002 (see Ivanchik et al. 2005). The data used by Ivanchik were retrieved from the VLT archive along with the MIDAS based UVES pipline reduction procedures.
Additional observational data of QSO 0347-383 acquired in 2002 at the same telescope but not previously analyzed[3] is taken into account here.
Paolo Molaro from the Osservatorio Astronomico di Trieste carefully reduced the overlooked dataset again to meet present requirements and provided the data for analysis.
The UVES observations comprised of 6 × 80 minutes-exposures of QSO 0347-383 on several nights, thus adding another 28.800 sec of exposure time. The journal of these observations as well as additional information is reported in Table 3.1. Three UVES spectra were taken with the DIC1 and setting 390+580 nm and three spectra with DIC2 and setting 437+860, thus providing blue spectral ranges between 320-450 and 373-500 nm respectively. The spectrum of QSO 0347-383 has no flux below 3700 Å due to the Lyman discontinuity of the z_{abs}=3.023 absorption system. The slit width was set to 1″ for all observations providing a Resolving Power of ∼ 40.000. The seeing was varying in the range between 0.5″ to 1.4″ as measured by DIMM but normally seeing at the telescope is better than the value given by DIMM. The CCD pixels were binned by 2×2 providing an effective 0.027-0.030 Å pixel, or 2.25 kms^{-1} at 4000 Å

---

[2]Program ID 68.A-0106.
[3]Program ID 68.B-0115(A).

Table 3.1: Journal of the observations

Date	Time	λ	Exp(sec)	Seeing (arcsec)	airmass	S/N (mean)
2002-01-13	03:42:54	390	4800	1.7	1.5	20
2002-01-14	02:13:24	390	4800	1.0	1.2	28
2002-01-15	00:43:32	437	4800	0.96	1.0	67
2002-01-18	03:25:04	437	4800	1.63	1.4	49
2002-01-24	02:20:14	437	4800	1.07	1.7	29
2002-02-02	01:33:58	390	4800	0.5	1.2	37

along dispersion direction.

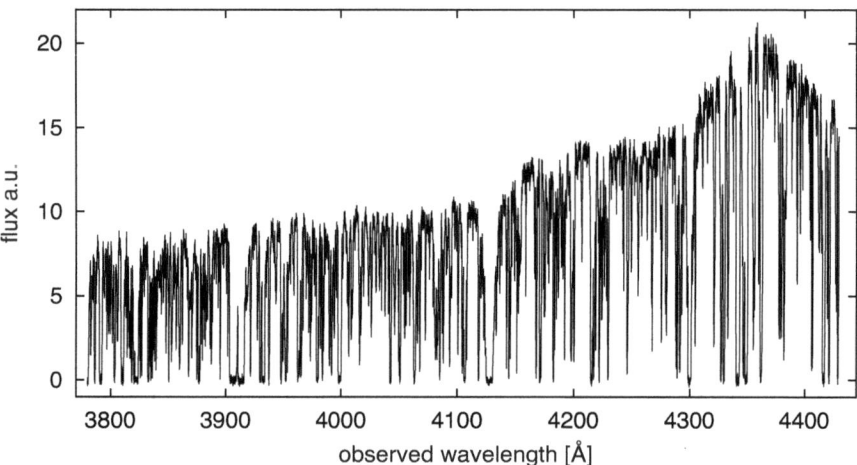

Figure 3.2: Co-added data of all 15 spectra of QSO 0347-383 (`CoAdd`). For the final analysis, however, the different spectra were not co-added but fitted simultaneously after correction for individual velocity shifts.

3.1.2 Reduction

The standard UVES pipeline has been followed for data reduction. This includes sky subtraction and optimal extraction of the spectrum. Typical residuals of the wavelength calibration were of ~ 0.5 mÅ or ~ 40 m s^{-1} at 4000 Å. The spectra were reduced to barycentric coordinates and air wavelengths have been transformed to vacuum by means of the dispersion formula given by Edlén (1966). Proper calibration and data reduction will be the key to detailed analysis of potential variations of fundamental constants. The influence of calibration issues on the data quality is hard to measure and the magnitude of the resulting systematic error is under discussion. The measurements rely on detecting a pattern of small relative wavelength shifts between different transitions spread throughout the spectrum. Normally, quasar spectra are calibrated by comparison with spectra of a hollow cathode thorium lamp rich in unresolved spectral lines. However several factors are affecting the quality of the wavelength scale. The paths for ThAr light and quasar light through the spectrograph are not identical thus introducing small distortions between ThAr and quasar wavelength scales. In particular differences in the slit illuminations are not traced by the calibration lamp. Since source centering into the slit is varying from one exposure to another an offset in the zero point of the scales of different frames is induced which could be up to few hundred of $\mathrm{m\,s^{-1}}$. In section 3.2.2 an estimate of these offsets which result in a mean offset of 168 $\mathrm{m\,s^{-1}}$ are provided as well as a procedure to avoid this problem. Laboratory wavelengths are known with limited precision which is varying from line to line from about 15 $\mathrm{m\,s^{-1}}$ for the better known lines to more than 100 $\mathrm{m\,s^{-1}}$ for the more poorly known lines (Murphy et al. 2008; Thompson et al. 2009a). However, this is the error which is reflected in the size of the residuals of the wavelength calibration.

Effects of this kind have been investigated at the Keck/HIRES spectrograph by comparing the ThAr wavelength scale with one established from I2-cell observations of a bright quasar by Griest et al. (2010). They found both absolute and relative wavelength offsets in the Keck data reduction pipeline which can be as large as 500 - 1000 $\mathrm{m\,s^{-1}}$ for the observed wavelength range. Such errors would correspond to $\Delta\lambda \sim 10-20\,\mathrm{m\AA}$ and exceed by one order of magnitude presently quoted errors (Thompson et al. 2009a). Examination of the UVES spectrograph at the VLT carried out via solar spectra reflected on asteroids with known radial velocity showed no such dramatic offsets being less than $\sim 100\ \mathrm{m\,s^{-1}}$ (Molaro et al. 2008a) but systematic errors at the level of few hundred $\mathrm{m\,s^{-1}}$ have been revealed also in the UVES data by comparison of relative shifts of lines with comparable response to changes of fundamental constants (Centurión et al. 2009). These examples well show that current $\Delta\mu/\mu$-analysis based on quasar absorption spectra at the level of a few ppm enters the regime of calibration induced systematic errors. While awaiting a new generation of laser-comb-frequency calibration, today's efforts to investigate potential variation of fundamental physical constants require true consideration of the strong systematics.

The additional observations considered here were originally taken for other purposes and the ThAr lamps are taken during daytime, which means several hours before the science exposures and likely under different thermal and pressure conditions. However, in this thesis the possibility of different zero points of the individual images is bypassed via the rare case of independent observations. Instead of co-adding all the spectra, first the global velocity shifts between the spectra is computed with the procedure described in the following section and also the whole uncertainties coming from the wavelength accuracies are utilized as part of the analysis procedure.

3.2 Preprocessing of data

3.2.1 Conditioning of flux

The UVES data reduction procedure delivers the error spectrum along the optimally extracted spectrum. The given error in flux of all 15 spectra was tested against the zero level noise in saturated areas. A broad region of saturated absorption is available near 3906 Å in the observers frame. Figure 3.3 displays the wavelength range in question for the co-added data. The underlying error was of course derived from the 15 individual spectra. Statistical analysis revealed a variance corresponding to $\sim 120\%$ of the given error on average for the 15 spectra (`ErrMeter`).

This means that normally errors that rely to the standard extracted routine are probably underestimated by a comparable amount. In particular the standard deviation of the flux between 3903.8 Å and 3908.7 Å (roughly 160 samples) was compared with the average of the specified error for that range. In this analysis for each of the spectra the calculated correction factor was applied to all samples.

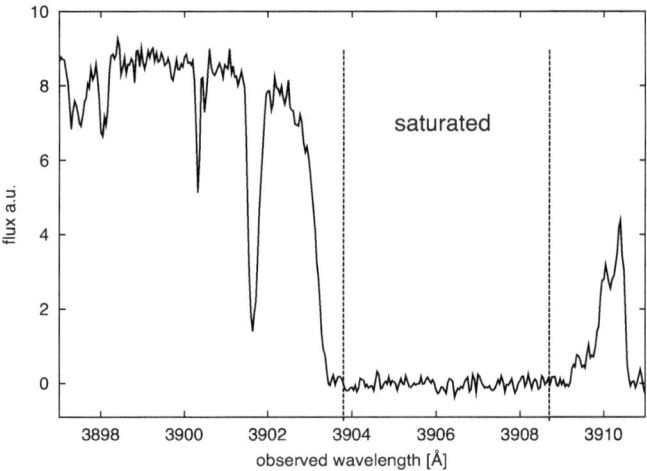

Figure 3.3: Range of saturated absorption in the spectrum of QSO 0347-383 that can be utilized to determine the minimal present gaussian error of the data. Plotted for the co-added data for illustration purposes.

3.2.2 Correction for individual shifts

Prior to further data processing the reduced spectra are reviewed in detail. Es described in more detail in section 3.1.1, the first data set (henceforward referred to as set A) consists of nine separate spectra observed between 7th and 9th of January in 2002 (see Ivanchik et al. 2005). The second set of 6 spectra (B) was obtained between January 13th and February 2nd in 2002 (see Table 3.1).

Due to slit illumination effects and grating motions the individual spectra are subject to small shifts – commonly on sub-pixel level – in wavelength. These shifts are particularly crucial in the process of co-addition of several exposures. To estimate these shifts all spectra were interpolated by a polynomial using *Neville's algorithm* to conserve the local flux (see Fig. 3.4). Neville's algorithm is based on the Newton form of the interpolating polynomial and the recursion relation for the divided differences. The interpolating polynomial of degree $N-1$ through the N points $y_0 = f(x_0), y_1 = f(x_1), \ldots, y_{N-1} = f(x_{N-1})$ is given by Lagrange's classical formula,

$$P(x) = \frac{(x-x_1)(x-x_2)\ldots(x-x_{N-1})}{(x_0-x_1)(x_0-x_2)\ldots(x_0-x_{N-1})} y_0$$
$$+ \frac{(x-x_1)(x-x_2)\ldots(x-x_{N-1})}{(x_1-x_0)(x_1-x_2)\ldots(x_1-x_{N-1})} y_1 + \ldots \quad (3.1)$$
$$+ \frac{(x-x_1)(x-x_2)\ldots(x-x_{N-1})}{(x_{N-1}-x_0)(x_{N-1}-x_1)\ldots(x_{N-1}-x_{N-2})} y_{N-1}.$$

There are N terms, each a polynomial of degree $N-1$ and each constructed to be zero at all of the x_i except one, at which it is constructed to be y_i. Instead of implementing the Lagrange formula directly, Neville's algorithm was used which proceeds by first fitting a polynomial of degree 0 through the point (x_k, y_k) for $k = 1, \ldots, n$, e.g., $P_k(x) = y_k$. A second iteration is then performed in which P_i and P_{i+1} are combined to fit through pairs of points, yielding P_{12}, $P_{23}, \ldots$. The procedure is repeated, generating a "pyramid" of approximations until the final result is reached. For example, with $N=4$:

$$\begin{array}{cccccc}
x_1: & y_1 = P_1 & & & & \\
 & & P_{12} & & & \\
x_2: & y_2 = P_2 & & P_{123} & & \\
 & & P_{23} & & P_{1234} & \\
x_3: & y_3 = P_3 & & P_{234} & & \\
 & & P_{34} & & & \\
x_4: & y_4 = P_4 & & & &
\end{array}$$

Neville's algorithm is a recursive way of filling the numbers in the tableau a column at a time, from left to right. It is based on the relationship between a "daughter" P and its two "parents". The final result can then be expressed as:

$$P_{i(i+1)\ldots(i+m)} = \frac{(x-x_{i+m})P_{i(i+1)\ldots(i+m-1)}}{x_i - x_{i+m}} + \frac{(x_i - x)P_{(i+1)(i+2)\ldots(i+m)}}{x_i - x_{i+m}} \quad (3.2)$$

This recurrence works since the two parents already agree at points $x_{i+1} \ldots x_{i+m-1}$. Equation 3.2 was implemented from scratch in the programming language C to obtain a tolerable execution speed in comparison to interpreters such as IDL or MIDAS (used in `ShiftCheck`).

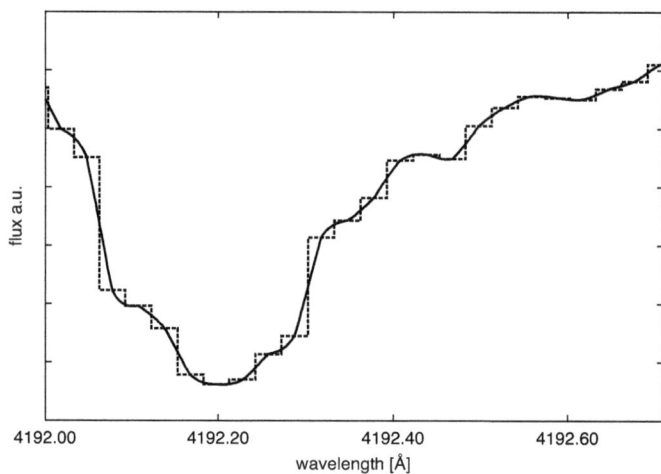

Figure 3.4: The original flux (*dashed steps*) is interpolated by a polynomial using *Neville's algorithm* (*solid line*) to conserve the local flux.

The resulting pixel step on average is 1/20 of the original data. Each spectrum was compared to the remaining 14 spectra. For each data point in a spectrum the pixel with the closest wavelength was taken from a second spectrum. Their deviation in flux was divided by the quadratic mean of their given errors in flux. This procedure was carried out for all pixels inside certain selected wavelength intervals.

Only certain wavelength ranges are taken into account to avoid areas heavily influenced by cosmic events or areas close to overlapping orders, resulting in a mean deviation of two spectra. The second spectrum is then shifted against the first one in steps of ~ 1.5 mÅ, according to the binsize of the subsampled spectrum. Since for each inspected shift, every data point can be compared independently, this routine was implemented using C and OpenMP to parallelize the process. The distribution of the considered wavelength intervals among the contributing processors is very helpful since each data set is enlarged by the above mentioned factor of 20 (`ShiftCheck`).

The run of the discrepancy of two spectra is of parabolic nature with a minimum at the relative shift with the best agreement. Fig. 3.5 shows the resulting curve with a parabolic fit. In this exemplary case the second spectrum shows a shift of 6.2 mÅ in relation to the reference spectrum. The clean parabolic shape verifies the approach. Table 3.2 shows the corresponding offsets for the 15 spectra. The offsets between the exposures are relevant with a peak to peak excursion up to almost 800 m s^{-1}. The average deviation is 2.3 mÅ or 170 m s^{-1} at 4000 Å. For further analysis in this paper all the 15 spectra are shifted to their common mean, which is taken as a reference position.

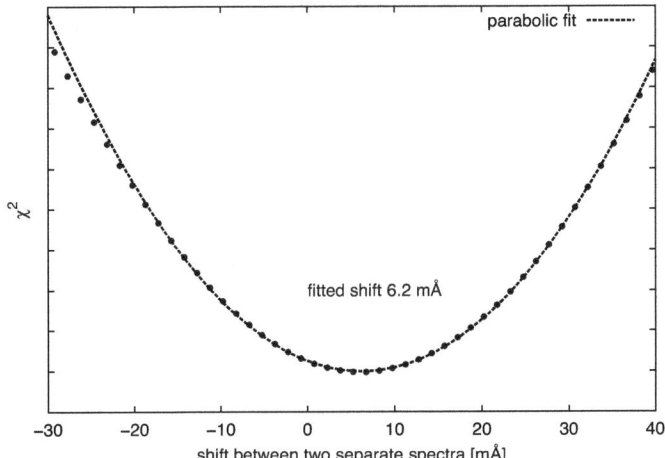

Figure 3.5: Exemplary plot of the sub-pixel cross-correlation. The resulting shift is ascertained via parabolic fit. In this case the two spectra are in best agreement with a relative shift of 6.2 mÅ or 0.465 kms^{-1}, respectively.

Section 5.4 illustrates its influence on the data analysis with respect to the previous analysis of the data set A, which have not considered this effect.

3.2.3 Selection of H$_2$ lines

The selection of suitable H$_2$ features for the final analysis is rather subjective. As a matter of course all research groups cross-checked their choice of lines for unresolved blends or saturation effects. The decision whether a line was excluded due to continuum contamination or not, however, relied mainly on the expert knowledge of the researcher and was only partially reconfirmed by the estimated uncertainty of the final fitting procedure. This thesis puts forward a more generic approach adapted to the fact that two distinct observations of the same object are available. Each H$_2$ signature is fitted with a single component. The surrounding flux is modelled by a polynomial and the continuum is rectified accordingly (see section 3.3.2). A selection of 52 (in comparison with 68 lines for that system by King et al. (2008) lines is fitted separately for each dataset of 9 (A) and 6 (B) exposures, respectively. In this selection merely blends readily identifiable or emerging from equivalent width analysis are excluded. The visual impression of the quality of the signature in terms of strength and environment is not the decisive factor. See section 6.2.2 on page 71 for further details on the identification of H$_2$ signatures.

Each rotational level is fitted with conjoined line parameters except for the redshift naturally. The data are not co-added but analyzed simultaneously via the fitting procedure applied by

Table 3.2: Relative shifts of the observed spectra to their common mean. Spectra A1-A9 correspond to the observations of Program ID 68.A-0106, spectra B1-B6 to Program ID 68.B-0115(A), respectively.

Spectrum	shift to mean [kms^{-1}]
A1	-0.203
A2	-0.135
A3	0.116
A4	-0.061
A5	0.268
A6	-0.031
A7	-0.249
A8	0.065
A9	0.249
B1	-0.084
B2	0.496
B3	0.039
B4	-0.339
B5	0.030
B6	-0.158
average deviation	0.168

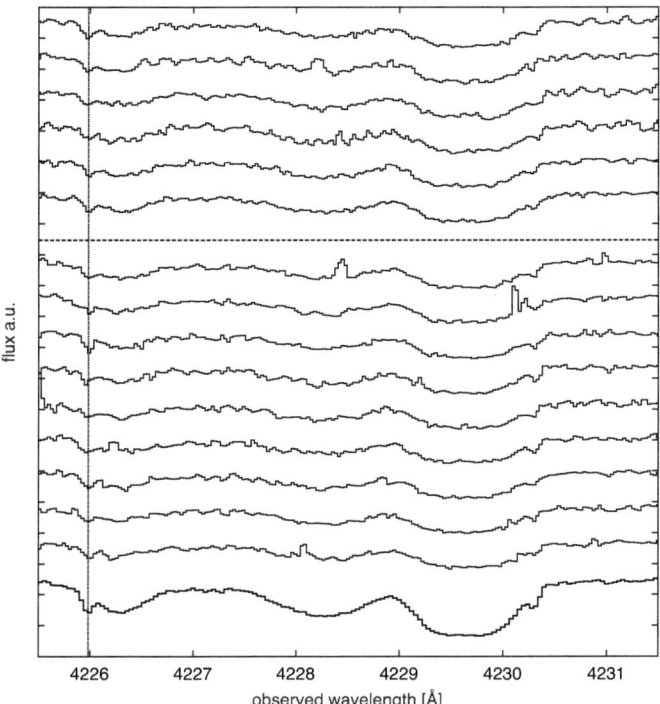

Figure 3.6: The 6 single spectra of set B (*top*), the 9 spectra of set A (*below*) separated by the *slashed line* and (not to scale) the corresponding co-added data (*bottom*) are plotted around the region of L4R1 (*vertical line*).

Quast et al. (2005).

For each of the 52 lines there are two resulting fitted redshifts or observed wavelengths, respectively, with their error estimates. To avoid false confidence, the single lines are not judged by their error estimate but by their difference in wavelength between the two data sets in relation to the combined error estimate. The absolute offset $\Delta\lambda_{\text{effective}}$ to each other is expressed in relation to their combined error given by the fit:

$$\Delta\lambda_{\sigma\Sigma1,2} = \frac{\Delta\lambda_{\text{effective}}}{\sqrt{\sigma_{\lambda_1}^2 + \sigma_{\lambda_2}^2}}. \tag{3.3}$$

Figure 3.7 reveals notable discrepancies between the two datasets, the disagreement is partially exceeding the $5\,\sigma$ level. Lines fitted with seemingly high precision and thus a low error reach higher offsets than lines with a larger estimated error at the same discrepancy in λ_{obs}. Clearly the lower error estimates merely reflects the statistical quality of the fit, not the true value of the specific line position.

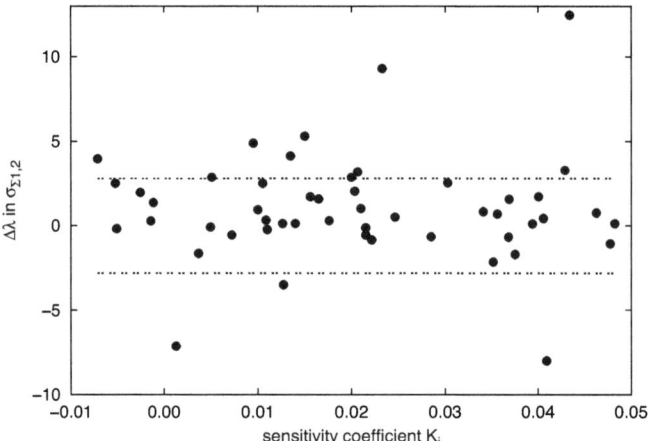

Figure 3.7: Selection of 52 apparently reasonable lines to be fitted separately for each dataset of 9 and 6 exposures, respectively. Their absolute offset $\Delta\lambda_{\text{effective}}$ to each other is expressed in relation to their combined error given by the fit (see Eq. 3.3). The dashed lines border the $3\,\sigma$ domain.

Since the fitting routine is known to provide proper error estimates (Quast et al. 2005; Wendt and Reimers 2008; Wendt et al. 2009), the dominating source of error in the determination of line positions is due to systematic errors. This result indicates calibration issues of some significance at this level of precision. The comparison of two independent observation runs reveals a source of error that cannot be estimated by the statistical quality of the fit alone. For further analysis only lines that differ by less than $3\,\sigma$ are taken into account.

This criterion is met by 36 lines. Fig. 3.8 shows three exemplary H_2 features corresponding to the transitions L5R1, L5P1, L5R2. All have similar sensitivity towards changes in μ. However, L5P1 fails the applied self consistency check between the two data sets and is excluded in the further analysis. Table 3.3 lists the excluded lines for the follow up analysis.

It is noteworthy that line selections of this absorption system by other groups diverge from each other by a large amount. King et al. (2008) processed a total of 68 lines. By reconstructing the continuum flux with additionally fitted lines of atomic hydrogen they felt confident not to care about the relative position of the H_2 features next to the Lyman-α forest. Fitting H_2 features as single lines, however, is affected by the surrounding flux and its nature as simulations have shown (Wendt and Reimers 2008).

Thompson et al. (2009a) selected 36 lines for analysis which differs from the semi-automatic choice of lines presented here by almost 40%. Different approaches, line selections and in the end applied methods contribute to a more solid constraint on variation of fundamental

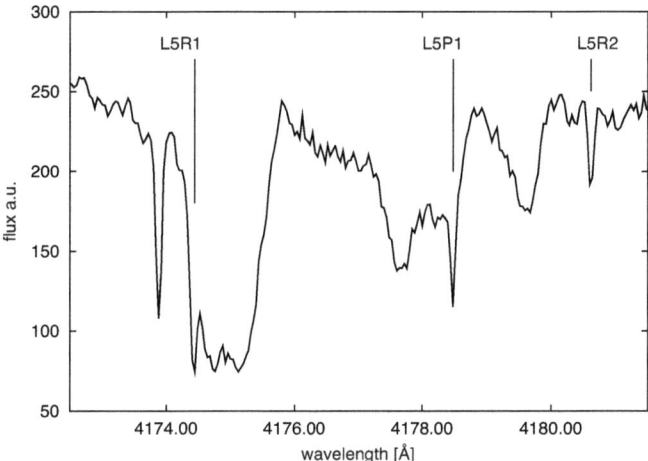

Figure 3.8: Part of the co-added observed spectrum near 4176 Å. The data however, were not co-added for the fit. L5R1 and L5R2 match the $3\,\sigma$ criterion, L5P1 does not and was hence excluded (see Table 3.3).

constants. This variety is mandatory to understand contradicting findings, not only in case of the proton-to-electron mass ratio. Table 4.1 reports the molecular line position and relative errors.

Figure 3.9 plots the observed redshifts of the lines with their estimated positioning error versus their corresponding sensitivity towards changes in μ. The datapoints with the given errorbars are from this analysis, the other points by Ubachs et al. (2007) and Thompson et al. (2009a), who published the individual fit parameters of their analysis.

The redshifts derived are of $z_{\text{abs}} = 3.0248969(56)$, $3.0248988(29)$ and $3.0248987(61)$ for this analysis, Ubachs et al. (2007) and Thompson et al. (2009a) respectively, which is not surprising being based at least partially on the same data. All three analyses are based on the same source for sensitivity coefficients as well, allowing the comparative plot.

The distribution of positioning errors for the mentioned works is illustrated in Figure 3.10. The three sets of measure show a significant scatter around the mean quite in excess of the error in line position which is suggestive of the presence of systematic errors.

The chosen $\Delta\lambda$ criterion for line selection permits evaluation of the self-consistency of a line positioning via fit for the involved data. While the availability of two independent observations on short time scale is rather special, it illustrates one applicable modality to avoid relying on the fitting apparatus alone.

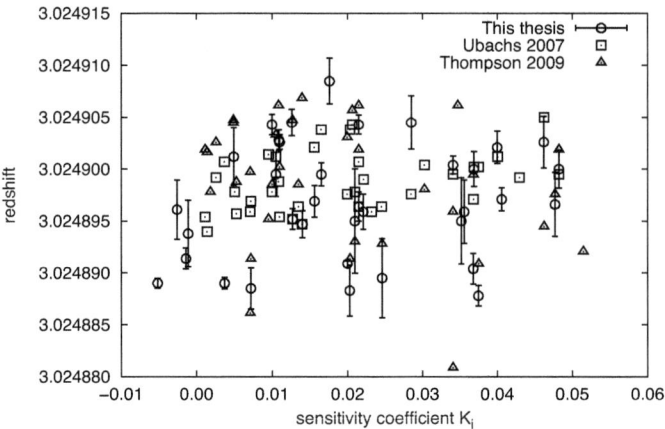

Figure 3.9: Final results in redshift vs. sensitivity coefficient K_i for this analysis (*circles*), Ubachs et al. (2007) (*squares*) and Thompson et al. (2009a) (*triangles*).

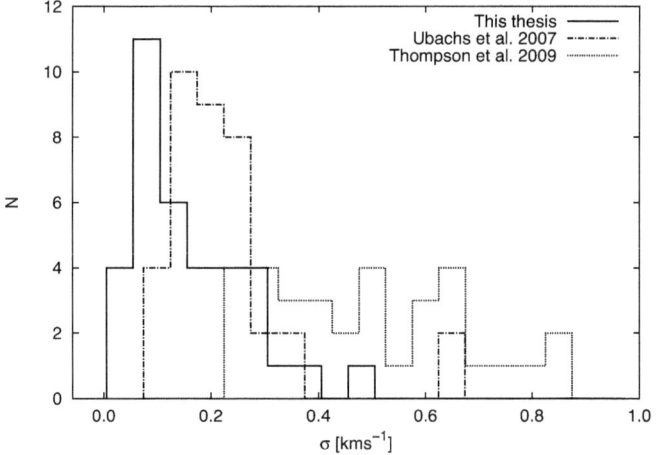

Figure 3.10: Line positioning errors in kms^{-1} for this thesis (*solid*), Ubachs et al. (2007) (*dashed*) and Thompson et al. (2009a) (*dotted*), binned to $50~\text{m s}^{-1}$.

Table 3.3: Excluded lines

Line ID	K factor	Lab. wavelength [Å]
L12P2	0.0434	966.2755
L11P1	0.0429	973.3345
L11P2	0.0409	975.3458
W0Q2	-0.0071	1010.9384
L7R1	0.0303	1013.4370
L6R2	0.0245	1026.5283
L6P2	0.0232	1028.1058
L5P1	0.0206	1038.1571
L5P2	0.0186	1040.3672
L4R2	0.0150	1051.4985
L4P2	0.0135	1053.2843
L3R2	0.0095	1064.9948
L2R1	0.0050	1077.6989
L2R3	0.0013	1081.7113

3.3 Fitting

Since the methods for fitting a theoretical line profile to observed data is manifold some possible procedures are described in more detail. As a first step an initial set of parameters is evaluated. Most fitting procedures depend on a reasonable manually selected set to bring the iterative algorithm on the right track so to speak. The most naive approach is to fit a line feature by a model based on the initial set and then vary each free parameter independently to find a minimum in the objective function describing the discrepancy between observation and fit.

This approach is of course unpractical, since the number of necessary calculations of the synthetic profile increases exponentially with each free parameter. Also the parameter range and resolution must be set sufficiently large to ensure a global minimum. Furthermore a reasonable range must be manually selected to assure that the covered parameter space does include the best-fit parameters.

Only very recently some groups experiment with exploration of the whole parameter space. King et al. (2008) for example implemented a Monte Carlo method to cover all possible combinations of fitting parameters. Unfortunately Monte Carlo methos scale exponentially with increasing dimensionality and are hence impractical for non-trivial situations. A slight improvement was achieved by the utilization of Markov Chain Monte Carlo simulations. The combination of Markov Chain methods and Monte Carlo simulation degrades merely polynomial with higher dimensionality but also introduces non-trivial correlation between samples and can in principle not be parallelized. Furthermore the underlying probability distribution has to be preassigned. It is therefor not feasible for realistic fitting tasks with todays computer power.

Even though it proves to be an valuable technique to estimate the statistical precision of a certain set of parameters, it cannot reveal anything about the accuracy of the data modelling and further does not produce traceable results. It should only be used as an supplemental method.

To avoid the need to inspect the whole parameter space, the common approach is to collect information about the local topology of the objective function by calculating its partial derivatives for each free parameter. This ensures a far more rapid convergence to a nearby minimum. This method (as implemented in the Levenberg-Marquardt algorithm for example) has the deficiency of relying on the initial parameter set, since in a straight forward implementation of this algorithm, possibly only a local minimum is found – depending on the situation not necessarily the global minimum. The first iteration step is based on the initially selected parameter set. Afterwards a second set of parameters is evaluated. The Levenberg-Marquardt algorithm interpolates a gradient of χ^2 in respect to the free model parameters and thereby ascertains the "direction" in the parameter space towards the local minimum. The second derivative of χ^2, more generally the Hessian matrix[4] delivers the appropriate stepsize for each parameter. Depending on the implementation, this stepsize is scaled additionally to match the required

---

[4]The Hessian matrix is the square matrix of second partial derivatives of a scalar-valued function.

needs. A new parameter set based on the information of the direction and the selected stepsize is evaluated and in case of a smaller χ^2 value used as fulcrum for the next iterationstep. The additional scaling is used to increase the resolution in the parameterspace with each iteration step. The iteration is completed when the change between successive χ^2 values falls below a certain threshold value. The value of the minimal χ^2 depends on the number of effective degrees of freedom. A value independent on the number of data points or the selected model is described by the normalized χ^2:

$$\chi^2_{\text{norm}} = \frac{\chi^2}{r}. \tag{3.4}$$

In the ideal case the final value of χ^2_{norm} reaches unity. That would mean that the total deviation between model and data equals the measurement errors of the data. A value below 1 thus indicates an invalid model or overestimated individual errors σ_i.

The approach applying evolutionary algorithms, such as the code implemented by Quast et al. (2005) that is used here, is based on stochastics (see, e.g., Hansen and Ostermeier 2001). The principle is similar to the one described above but instead of a manual first guess of initial parameters, several sets of random initial parameters are computed automatically over a parameter range that merely needs rough preselection. The most successful of those build the centres of other groups of random, yet less wide spread sets of parameters and so forth. This stochastic approach can additionally be fine- tuned by adjusting the expansiveness of each successive group of random parameters in both the parameter space and in quantity. With a sufficiently widespread cluster of parameter sets, theoretically the global minimum will always be reached. In practice a compromise must be found between computing time and success in reaching the global minimum. Its drawback is of course its inefficiency in terms of computing power and the need to check up the final fit on physically reasonable parameters. However, the principle of evaluating multiple groups of parametersets independently of each other allows for consequent parallel computing.

The element of randomness requires a full completion of the iterative process. Whilst the Levenberg-Marquardt algorithm usually converges rather soon towards the final parameters the evolutionary procedure can only give reliable results after all branches of parameter groups are evaluated. Some principle problems of redundancy in parameter sets for complex models have yet to be overcome as well. In praxis two or multiple line profiles whose parameter spaces significantly overlap can cause the algorithm to lock up on rare occasions. The dynamical scaling towards regions with small Hessian matrix to gain higher resolution is implemented by an increasingly smaller scattering of random parameter sets for successive child populations of parameter sets.

This fitting program uses a simplified pseudo-Voigt-function to generate the synthetic line profiles. For weak lines a mere gaussian profile would suffice since natural line broadening has no noticeable impact on the line shapes.

The program of Quast et al. (2005) was written in C++ under the application of OpenMP to parallelize the fitting procedure and distribute the computation of the individual parameter sets among the available CPU cores. The problem scales very well and the accumulated time for a complete fit in a multitude of spectra is justifiable.

Merely with minor adjustments mainly in the input/output interface, the fitting procedure could be adjusted to the special requirements when probing a variation in μ (`RQFit`).

The limitation of OpenMP, one single shared memory for all processes working in parallel plays no major role since the computations though large in number are comparably simple in detail and do not require notably large amounts of memory.

3.3.1 Simultaneous fit vs. co-added fit

The fitting program as described in the preceding section has the ability to fit several lines simultaneously to one subset of free parameters. Individual parameters can be cross referenced to each other and the code is under the constraint to find one best joint fit for these parameters. This feature is of great convenience since several line parameters as broadening width or column density are usually shared by a whole group of lines and do not differ among each other. This concept has even been expanded to fit several separate spectra simultaneously. So even for overlapping wavelength regions, the continuum and for example the redshift are fitted locally but the line width is fitted to all participant spectra and lines.

This method appears to be superior to a rough prior co-adding of spectra. Many errors coarse co-addition might introduce can be avoided by this way, since there is no need to rebin the data. Usually, rebinning leads to even-spaced data points and the flux is redistributed among the single bins. This degrades the original alignment of the reduced data, which – for geometric reasons concerning the projection onto the CCD chip and later reductions steps like the vacuum-air correction – has a differential gradient of spacing. See, e.g., Levshakov et al. (2002) for a discussion of this effect.

A rebin would introduce a stronger correlation of pixels. The impact of such correlations on the results is not well-known (see, e.g., Aitken 1934). A χ^2 minimization procedure does rely on independent data points though. The concept of simultaneous fit has the advantage of taking physical conditions into account in form of further restrictions on the degrees of freedom.

3.3.2 Continuum handling

An accurate estimate of the true continuum of a spectrum is essential for a good fit of an absorption line. Especially for optically thin lines the determined column density is very sensitive to variations in the continuum fit. Particularly in damped Lyman-α systems (DLA) at large redshifts there hardly is any unblended continuum detectable in the range of H_2 absorption due to the Lyman-α forest. There are numerous techniques and strategies to gain a reliable estimation of the true continuum despite its contamination.

A rather generic approach is specific to pure absorption spectra and assuming that the data points with the highest photon count contribute to the continuum, since there are no emission features. Therefore a polynomial function of low order is fitted iteratively to a selected subset of data points above a certain threshold. This threshold is increased or rather the range between Flux_{min} and Flux_{max} (if there are some spikes or cosmics in the spectra) is narrowed with each iteration step until a minimum variance or the predefined minimum number of selected data points is reached. This method is only applicable though when a reasonable amount of undisturbed continuum emission is present in the wavelength range of interest and hence not applicable to the spectra at hand.

Another way of dealing with the continuum is to construct it manually. The curve of the estimated continuum is drawn or based on some interpolation points. The continuum level simply is set at the upper end of an absorption line feature to disregard all influences of other lines nearby. Dealing with the continuum in this manner yields a great uncertainty that only a curve of growth analysis can reveal. Albeit this is one of the few options to deal with heavily contaminated and partially blended lines at all and was carried out by Ivanchik et al. (2002) for example.

A refinement of the latter approach is to not only fit the one line of interest and border it by an artificial continuum but to fit a series of lines with free parameters to match the disturbed measured flux. The actual line of interest is then embedded into a group of lines that are used to fit the surrounding flux level. The true continuum, however, needs to be estimated but for a larger less alternating environment. In some cases the extra lines may disturb the shape of the measured main component and this procedure needs special supervision.

In this work, the continuum is fitted by the code together with the lines. The fit is based on the flux adjoining the line feature. The accuracy of the continuum fit increases significantly by the option of simultaneous fitting. All molecular hydrogen lines observed at the same rotational level should have identical physical properties, as in column density and line width. This constrains the uncertainty of the continuum since the resulting column density has to be consistent with the simultaneous fits of the more secure continuum regions and the identified H_2 absorption features are spread over the full range of the observed spectrum. However, a mere χ^2 consideration to evaluate the quality of the fit is invalid, since the model of a single line does not describe the observed flux sufficiently.

In cases of a continuum flux with apparent influences of broad Lyman-α features or general contamination, a parabolic or cubic function was used to fit the background to the observed flux. See Figure 3.11 of L3R3 and L6R3 for an exemplar of a parabolic fit. These exemplary lines already represent the extreme end of non-constant bordering continuum flux.

An additional issue not yet discussed in connection with varying μ are the oscillator strengths. Whereas the transition frequencies have been revised several times over the last few years with improved instrumentation and numerical methods (see section 2.2.3), the latest data for the oscillator strength of molecular hydrogen are still from tables by Abgrall et al. (1993a,b). UV-laser experiments are not yet capable of arranging the required measurements: "The oscillator strengths can only be derived by calculations based on coarse approximations. Verification

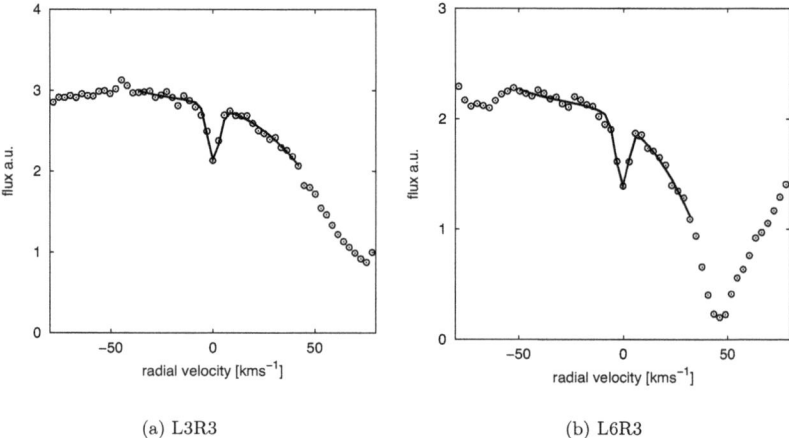

(a) L3R3 (b) L6R3

Figure 3.11: Exemplary cases of continuum matching via parabolic fit to the observed flux for the Lyman lines L3R3 and L6R3.

of the listed values of the oscillator strengths by experimental methods appears impossible at laboratory conditions at the present state." (Ubachs, private communication).

Uncertainties in the oscillator strengths are not directly reflected in the measured line positions and hence the obtained redshift but they have influence on the quality of fit and the determination of the continuum. In particular the approach to fit a designated number of arbitrary line components in the vicinity of H_2 profiles is in particular error-prone to wrong oscillator strengths and therefore wrong equivalent widths.

Groups that apply this multi-component fit (see, e.g., King et al. 2008), in general add further components to a fit-region around a H_2 signature, until the lowest χ^2 value is reached. Alternatively, the residuals are inspected for remaining flux beyond the noise level.

This approach is vulnerable to wrong oscillator strengths and in the same way to underestimated errors of the observed flux (see section 3.2.1 for more details).

4 Results I

4.1 Determination of $\Delta\mu/\mu$

For the final analysis the selected 36 lines are fitted in all 15 shifted, error-scaled spectra simultaneously (see section 3.2.3 for details on the line selection criteria). The result of an unweighted linear fit corresponds to

$$\Delta\mu/\mu = (15 \pm 16) \times 10^{-6}, \tag{4.1}$$

at $z_{\text{abs}} = 3.025$. The stated finding of Eq. 4.1 can be translated into a rate of change. When imposing the assumptions of a linear cosmological expansion model, the redshift z is related to the look-back time T via:

$$T = T_0 \left[1 - (1+z)^{-3/2}\right], \tag{4.2}$$

with T_0 the age of the universe, which may be set at 13.7 Gyrs. For the quasar under consideration this yields a look-back time of ~ 12 Gyrs. With the further assumption that μ has varied linearly over time this corresponds to a rate of change of $\mathrm{d}\ln\mu/\mathrm{d}t = -1.2 \times 10^{15}\,\text{yr}^{-1}$. The negative sign must be interpreted as a decrease in μ over time.

This conversion into a rate is merely exemplary, however. The findings of this analysis yield no indication of variation with the current limit of accuracy.

Figure 4.1 shows the resulting plot. The complete list of lines is shown in Table 4.1.

The approach to apply an unweighted fit is a consequence of the unknown nature of the prominent systematics. Uncertainties in wavelength calibration cannot be expressed directly as an individual error per line. The graphed scatter in redshift can not be explained by the given positioning errors alone. The likeliness of the data with the attributed error being linearly correlated is practically zero. The fit to the data is not self-consistent. For this work the calibration errors and the influence of unresolved blends are assumed to be dominant in comparison to individual fitting uncertainties per feature.

For the following analysis the same error is adopted for each line. With an uncertainty in redshift of 1×10^{-6} we obtain: $\Delta\mu/\mu = (15 \pm 6) \times 10^{-6}$. However the goodness-of-fit is below 1 ppm and is not self consistent. Judging by that and Fig. 4.1, a reasonable error in observed redshift should at least be in the order of 4×10^{-6}. The weighted fit gives: $\Delta\mu/\mu = (15 \pm 14) \times 10^{-6}$. This approach is motivated by the goodness-of-fit test:

$Q(\chi^2|\nu)$ is the probability that the observed chi-square will exceed the value χ^2 by chance even

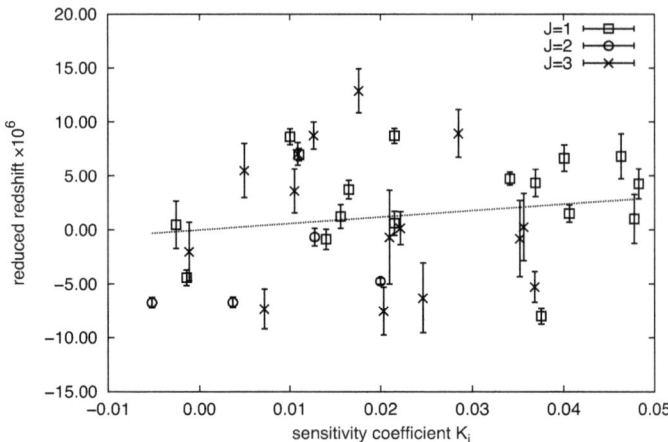

Figure 4.1: The unweighted fit to the measured redshifts of the H_2 components in QSO 0347-383 corresponds to $\Delta\mu/\mu = (15 \pm 16) \times 10^{-6}$. The error bars represent merely to the fitting uncertainty in the order of 180 m s^{-1} on average. Note, that at such a high scatter $z_{Ki=0}$ differs from $\bar{z}$ by less than 1 σ_z.

for a correct model, ν is the number of degrees of freedom. Given in relation to the incomplete gamma function:

$$Q(\chi^2|\nu) = \Gamma\left(\frac{\nu}{2}, \frac{\chi^2}{2}\right). \tag{4.3}$$

Assuming a gaussian error distribution, Q gives a quantitative measure or the goodness-of-fit of the model. If Q ist very small for some particular data set, then the apparent discrepancies are unlikely to be chance fluctuations. More probable is either that the model is wrong or the size of the measurement errors is larger than stated. However, the chi-square probability Q does not directly measure the credibility of the assumption that the measurement errors are normally distributed. In general, models with $Q < 0.001$ can be considered unacceptable. In this case the model is given and hence the low probability is due to underestimated errors in the data. Solely for given errors of ~ 300 m s^{-1}, corresponding to $\sim 4 \times 10^{-6}$ in redshift for QSO 0347-383 the goodness-of-fit parameter Q exceeds 0.001. The scale of the error appears to be ~ 300 m s^{-1} to achieve a self-consistent fit to the data.

For the data on QSO 0347-383 this corresponds to an error in the observed wavelength of roughly 4 mÅ, which is notably larger than the estimated errors for the individual line fits which ranges from 0.5 mÅ to 6.5 mÅ with an average of 2.5 mÅ (~ 180 m s^{-1}). The systematic error contributes an uncertainty of about 2 mÅ on average. The immediate calibration errors are in the order of 50 m s^{-1} for set B and presumably slightly larger for set A (see section 3.1.1).

Figure 4.2 plots the data with errorbars corresponding to 180 m s^{-1} and the total of 300 m s^{-1}.

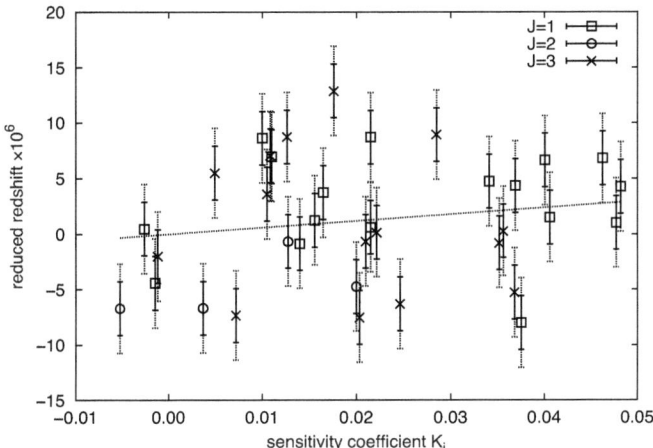

Figure 4.2: The data points are identical to Fig. 4.1. Here, the error bars represent the average positioning error (*solid*) and the additional systematic error (*dotted*) of $\sim 180\,\text{m s}^{-1}$ and $\sim 120\,\text{m s}^{-1}$, respectively.

The final result can be subdivided as:

$$\Delta\mu/\mu = \left(15 \pm (9_{\text{stat}} + 6_{\text{sys}})\right) \times 10^{-6}. \qquad (4.4)$$

The comparably high scatter in Figure 4.1 can partially be attributed to the approach to fit single H$_2$ components with a polynomial fit to the continuum. In special cases, contaminated flux bordering a H$_2$ signature can introduce additional uncertainty in positioning. Therefore checks for self-consistency and systematics are of utmost importance.

The determination of the different errors involved is on a par with the actual result. We believe that this result represents the limit of accuracy that can be reached with the given data set and the applied methods for analysis. The presented method yields a null result. The recent work by Thompson et al. (2009a) stated $\Delta\mu/\mu = (-28 \pm 16) \times 10^{-6}$ for a weighted fit based on the same system in QSO 0347-383. The stated errors in that work reflect the statistical uncertainties alone.

Note, that the given systematics of 2.7ppm for Keck/HIRES data given in Malec et al. (2010) are in first approximation estimated by the observed $\sim 500\,\text{m s}^{-1}$ peak-to-peak intra-order value reduced according to the number of molecular transitions observed, e.g. $\sim 500\,\text{m s}^{-1}/\sqrt{93} \sim 52\,\text{m s}^{-1}$.

The approach in Malec et al. (2010) neglects the very nature of systematic errors as they present an absolute limit to the achievable accuracy and thus – in contrast to statistical errors – cannot be reduced by larger samples (see also section 5.2).

Table 4.1: QSO 0347-383 Line List

Line ID	K_i	$\lambda_{\rm obs}$ [Å]	$\sigma_{\lambda_{\rm obs}}$ [Å]	$\lambda_{\rm lab}$ [Å]	$\sigma_{\lambda_{\rm obs}}$ [kms^{-1}]	$z_{\rm abs}$
L14R1	0.0462	3811.5038	0.0031	946.9804	0.247	3.0249025
W3Q1	0.0215	3813.2825	0.0012	947.4219	0.091	3.0249043
W3P3	0.0210	3830.3795	0.0064	951.6719	0.499	3.0248950
L13R1	0.0482	3844.0442	0.0023	955.0658	0.181	3.0248999
L13P1	0.0477	3846.6271	0.0039	955.7083	0.306	3.0248966
W2Q1	0.0140	3888.4352	0.0017	966.0961	0.128	3.0248948
W2Q2	0.0127	3893.2050	0.0013	967.2811	0.099	3.0248951
L12R3	0.0368	3894.7939	0.0019	967.6770	0.149	3.0248904
W2Q3	0.0109	3900.3288	0.0013	969.0492	0.097	3.0249028
L10R1	0.0406	3952.7477	0.0015	982.0742	0.110	3.0248972
L10P1	0.0400	3955.8160	0.0020	982.8353	0.154	3.0249022
L10R3	0.0356	3968.3977	0.0040	985.9628	0.302	3.0248960
L10P3	0.0352	3975.6657	0.0055	987.7688	0.412	3.0248950
W1Q2	0.0037	3976.4877	0.0007	987.9745	0.054	3.0248890
L9R1	0.0375	3992.7546	0.0013	992.0164	0.098	3.0248877
L9P1	0.0369	3995.9594	0.0022	992.8096	0.167	3.0249000
L8R1	0.0341	4034.7699	0.0011	1002.4521	0.085	3.0249004
L8P3	0.0285	4058.6575	0.0034	1008.3860	0.255	3.0249046
W0R2	-0.0052	4061.2132	0.0006	1009.0249	0.047	3.0248890
L7P3	0.0246	4103.3836	0.0052	1019.5022	0.379	3.0248894
L6R3	0.0221	4141.5640	0.0023	1028.9866	0.168	3.0248960
L6P3	0.0203	4150.4349	0.0034	1031.1926	0.245	3.0248882
L5R1	0.0215	4174.4204	0.0019	1037.1498	0.139	3.0248963
L5R2	0.0200	4180.6152	0.0005	1038.6903	0.034	3.0248910
L5R3	0.0176	4190.5690	0.0031	1041.1588	0.218	3.0249086
L4R1	0.0165	4225.9822	0.0016	1049.9597	0.111	3.0248994
L4P1	0.0156	4230.2974	0.0021	1051.0325	0.151	3.0248969
L4R3	0.0126	4242.1531	0.0018	1053.9761	0.126	3.0249045
L4P3	0.0105	4252.1911	0.0030	1056.4714	0.211	3.0248994
L3R1	0.0110	4280.3234	0.0010	1063.4601	0.071	3.0249027
L3P1	0.0100	4284.9349	0.0014	1064.6054	0.097	3.0249043
L3R3	0.0072	4296.4822	0.0028	1067.4786	0.198	3.0248884
L3P3	0.0049	4307.2114	0.0040	1070.1409	0.276	3.0249012
L2P3	-0.0011	4365.2399	0.0046	1084.5603	0.318	3.0248937
L1R1	-0.0014	4398.1291	0.0015	1092.7324	0.100	3.0248913
L1P1	-0.0026	4403.4456	0.0042	1094.0520	0.283	3.0248961
average			0.0025		0.184	

4.2 Result via discrete line pairs

To accomplish a robust bound on the variation of μ, additional alternative approaches are advisable. $\Delta\mu/\mu$ can also be obtained by using merely two lines that show different sensitivity towards changes in the proton-to-electron mass ratio.

The set of derived redshifts with their corresponding sensitivity coefficients is sorted and pairs of two lines are selected via complete permutation. Line pairs that differ in sensitivity below a certain threshold were rejected (`ToPair`).

Another criterion is their separation in the wavelength frame to avoid pairs of lines from different ends of the spectrum and hence in particular error-prone (see, e.g., Griest et al. 2010). Several tests showed that a separation of $\Delta\lambda \leq 110\,\text{Å}$ and a range of sensitivity coefficients K_1-$K_2 \geq 0.02$ produces stable results that do not change any further with more stringent criteria. Pairs that cross two neighboring orders ($\sim 50\,\text{Å}$) show no striking deviations either, which would have indicated distinct problems in the wavelength calibration.

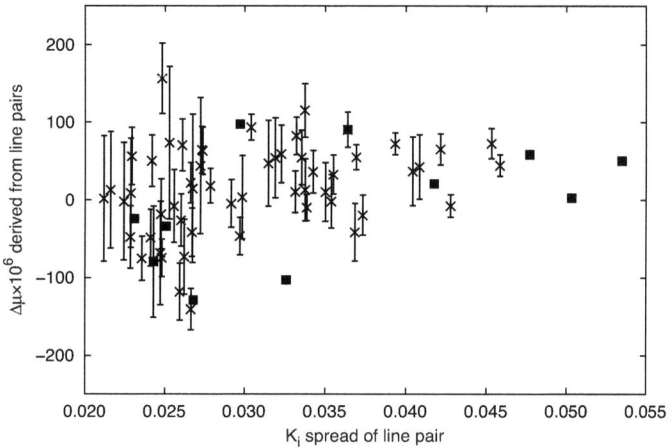

Figure 4.3: $\Delta\mu/\mu$ derived from individual line pairs (52) which are separated by less than 110 Å and show a difference in sensitivity of more then 0.02. The errorbars reflect the combined positioning error of the two contributing lines. The weighted fit corresponds to $\Delta\mu/\mu = (6 \pm 12) \times 10^{-6}$.

The *filled squares* graph 11 line pairs, selected to give the largest difference in sensitivity (≥ 0.02) towards variation in μ (See Table 4.2).

Figure 4.3 graphs the different values for $\Delta\mu/\mu$ derived from 52 line pairs that match the aforementioned criteria. Note, that a single observed line contributes to multiple pairs. Fitting of n line pairs that are sensitive to μ should be more precise than that based on a single pair.

However, the improvement is not as high as $1/\sqrt{n}$ because the different pairs are partially correlated via the redshift (and the sensitivity) of the same line that contributes to several pairs. If we consider n redshift differences $\{z_1 - z_0, z_2 - z_0, ..., z_n - z_0\}$, where z_0 is the reference redshift, then it is easy to show that the correlation coefficient $\kappa_{i,j}$ between two of them $(i \neq j)$ is given by

$$\kappa_{i,j} = \frac{1}{\sqrt{(1+s_i^2)(1+s_j^2)}}, \qquad (4.5)$$

where $s_i = \sigma_{z_i}/\sigma_{z_0}$ and $s_i = \sigma_{z_j}/\sigma_{z_0}$. Taking into account that the computational errors in determined redshifts, σ_{z_i}, are almost equal, we have $\kappa_{i,j} = \kappa \approx 1/2$.

The covariance matrix $Cov(z_i - z_0, z_j - z_0)$ contains n diagonal terms σ^2 and $n(n-1)$ non-diagonal terms $\kappa\sigma^2$, where σ^2 represents the variance in a single measurement. The error in the mean redshift caused by the positioning uncertainties (referred to as ε_{sys} hereafter) can be calculated as described by, e.g., Stuart and Ord (1994):

$$\varepsilon_{\text{sys}} = \left[\sum_{i=1}^{n}\sum_{j=1}^{n} \omega_i \omega_j Cov(z_i - z_0, z_j - z_0)\right]^{1/2}. \qquad (4.6)$$

In cases of equal accuracy, the weight $\omega_i = 1/n$ for each i. Then ε_{sys} is approximately equal to

$$\varepsilon_{\text{sys}} = \frac{\sigma}{n}\sqrt{n + n(n-1)\kappa} \approx \sigma\sqrt{\kappa}. \qquad (4.7)$$

Thus, the gain factor, $\sqrt{\kappa}$, is only about 0.7 for the line pairs in question. This alternative approach is in general not suited to provide a more precise result but it can verify the findings through an independent method. The discussed limit in precision through large number statistics applies for all kind of samples that hold some correlation.

The average value of the 52 redshifts-sensitivity ratios yields $\Delta\mu/\mu = 6 \pm 12 \times 10^{-6}$. The scatter is then related to uncertainties in the wavelength determination which is mostly due to calibration errors. The standard error is 8×10^{-6}.

The approach to use each observed line only once and thus avoid the mentioned correlation in the analysis is plotted in Fig. 4.3 as filled squares. The pairs to derive $\Delta\mu/\mu$ from were constructed by grouping the line with the highest sensitivity value together with the line corresponding to the lowest value for K_i and so on with the remaining lines. The distance in wavelength space between the two lines was no criterion and it ranges from 20Å to 590Å (see Table 4.2). Without reutilization of lines, 11 pairs with a coverage in sensitivity of $\Delta K_i \geq 0.02$ were found.

Evidently the usage of lines with comparably large distances in the spectrum has no influence on the results. This indicates that the contribution of global differential errors in the wavelength calibration appear to be small. Potential sources of such deviations would be the vacuum-air correction for example.

Table 4.2: Grouping all observed lines into 17 pairs of maximum K_i sensitivity not considering their separation in wavelength space (*rightmost* column).

Line 1	Line 2	$\Delta\mu/\mu$	ΔK_i	$\Delta\lambda$ [Å]
W0R2	L13R1	50.5 $\times 10^{-6}$	0.0535	-217.2
L1P1	L13P1	2.7 $\times 10^{-6}$	0.0503	-556.8
L1R1	L14R1	58.6 $\times 10^{-6}$	0.0477	-586.6
L2P3	L10R1	20.9 $\times 10^{-6}$	0.0417	-412.5
W1Q2	L10P1	90.4 $\times 10^{-6}$	0.0364	-20.7
L3P3	L9R1	-103.0 $\times 10^{-6}$	0.0326	-314.5
L3R3	L9P1	97.6 $\times 10^{-6}$	0.0297	-300.5
L3P1	L12R3	-128.9 $\times 10^{-6}$	0.0268	-390.1
L4P3	L10R3	-33.9 $\times 10^{-6}$	0.0251	-283.8
W2Q3	L10P3	-79.4 $\times 10^{-6}$	0.0243	75.3
L3R1	L8R1	-24.0 $\times 10^{-6}$	0.0231	-245.6
L4R3	L8P3	3.1 $\times 10^{-6}$	0.0159	-183.5
W2Q2	L7P3	-120.0 $\times 10^{-6}$	0.0119	210.2
W2Q1	L6R3	34.0 $\times 10^{-6}$	0.0082	253.1
L4P1	W3Q1	312.7 $\times 10^{-6}$	0.0059	-417.0
L4R1	L5R1	-154.6 $\times 10^{-6}$	0.0050	-51.6
L5R3	W3P3	-996.4 $\times 10^{-6}$	0.0034	-360.2

5 Error Analysis I

5.1 Quality of fit

A least-squares fit to the unweighted redshift versus sensitivity coefficient data can only yield the best fit to the data. It is not trivial to ascertain the quality of the fit (see chapter 4.1).

The uncertainties of the fit merely represent the certainty to which the found parameters describe indeed the best possible fit. It is not directly correlated to the distribution of the data itself and of course, it contains no information at all on the systematics underlying the data.

The weighted fit in principle is more suited to represent the quality of the data and its distribution but it is strongly biased by the uncertainty of the contributing errors. This is obvious for the present data with estimated positioning errors per line, that fail to reflect the scatter of the data.

A diverse approach to examine data is via the bootstrap method (see, e.g., Efron and Tibshirani 1986).

Bootstrapping is the practice of estimating properties of an estimator (such as its variance) by measuring those properties when sampling from an approximating distribution. One standard choice for an approximating distribution is the empirical distribution of the observed data. In the case where a set of observations can be assumed to be from an independent and identically distributed population, this can be implemented by constructing a number of resamples of the observed dataset (and of equal size to the observed dataset), each of which is obtained by random sampling with replacement from the original dataset.

For the data at hand this implies to select samples of 36 lines of the data set of observed lines. The linear fit to the measured redshift against the sensitivity data is applied to this new sample. This is redone for a large number of randomly selected samples. For 36 lines there are evidently 36^{36} possible samples, including all permutations of the same lines though. A few thousand samples are sufficient in general and Figure 5.1 shows the histogram of $\Delta\mu/\mu$ derived from 10.000 bootstrap samples.

Again the algorithm to perform the resampling and this large amount of least-square fits is implemented in C and OpenMP, since the individual fits are independent of each other and can easily be parallelized (`BootStrap`). The histogram is generated with a binning of 1×10^{-6} and Figure 5.1 also contains the fit of a gaussian (*solid line*) to the data. The distribution of the $\Delta\mu/\mu$ is slightly asymmetric but this has no impact on the findings.

The result of the bootstrap analysis corresponds to:

$$\Delta\mu/\mu = (14.9 \pm 12.0) \times 10^{-6}. \qquad (5.1)$$

This verifies the results and in particular the error estimation of section 4.1 that yielded $\Delta\mu/\mu = \left(15 \pm (9_{\text{stat}} + 6_{\text{sys}})\right) \times 10^{-6}$.

The FWHM of the bootstrap analysis incorporate the scatter of the data automatically and is in good agreement with the estimation of the systematic errors via goodness of fit tests. The FWHM of the Gaussian profile is:

$$\text{FWHM} = 2\sigma\sqrt{2\ln(2)}. \qquad (5.2)$$

The uncertainty corresponds to the inflection points of the gaussian curve and can be derived from the width of a fitted gaussian via $\sigma \sim \text{FWHM}/2.35$.

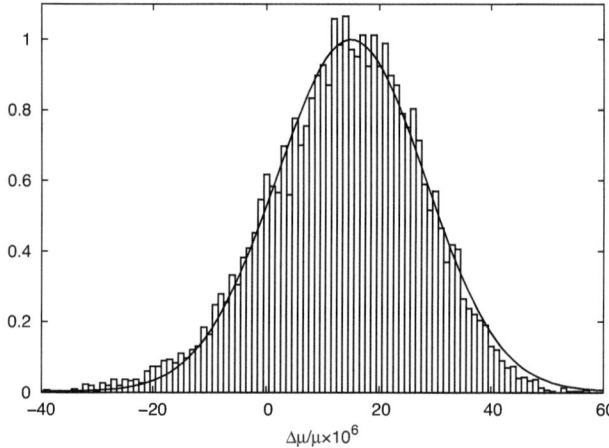

Figure 5.1: $\Delta\mu/\mu$ and in particular its uncertainty derived via 10.000 bootstrap samples. The gaussian fit yields a centroid at 14.9 and a FWHM of 28.2.

5.2 Standard Error

The terms "standard error" and "standard deviation" are often confused. The contrast between these two terms reflects the important distinction between data description and inference.

The standard deviation is a measure of variability. When we calculate the standard deviation of a sample, we are using it as an estimate of the variability of the population from which the sample was drawn. For data with a normal distribution, about 95% of individuals will have values within 2 standard deviations of the mean, the other 5% being equally scattered above and below these limits. Contrary to popular misconception, the standard deviation is a valid measure of variability regardless of the distribution. About 95% of observations of any distribution usually fall within the 2 standard deviation limits, though those outside may all be at one end.

When we calculate the sample mean we are usually interested not in the mean of this particular sample, but in the mean for individuals of this type-in statistical terms, of the population from which the sample comes. We usually collect data in order to generalize from them and so use the sample mean as an estimate of the mean for the whole population. Now the sample mean will vary from sample to sample; the way this variation occurs is described by the "sampling distribution" of the mean. We can estimate how much sample means will vary from the standard deviation of this sampling distribution, which we call the standard error (SE) of the estimate of the mean. As the standard error is a type of standard deviation, confusion is understandable. Another way of considering the standard error is as a measure of the precision of the sample mean.

The standard error of the sample mean depends on both the standard deviation and the sample size, by the simple relation $SE = SD/\sqrt{samplesize}$. The standard error falls as the sample size increases, as the extent of chance variation is reduced. By contrast the standard deviation will not tend to change with the size of our sample.

If we want to state how widely scattered some measurements are, we use the standard deviation. If we want to indicate the uncertainty around the estimate of the mean measurement, we quote the standard error of the mean.

Hence, the standard error is not suited to describe the quality of a data set, it merely represents the precision of a particular fit for example, not its accuracy.

5.3 Uncertainties in the sensitivity coefficients

At the current level of precision, the influence of uncertainties in the sensitivity coefficients K_i is minimal. It will be of increasing importance though when wavelength calibration can be improved by pedantic demands on future observations. Eventually Laser Frequency Comb calibration will allow for practically arbitrary precision and uncertainties in the calculations of sensitivities will play a role. The wavelength data for molecular hydrogen available is adequate even for the next generation telescopes, like the European Extremely Large Telescope (ELT) currently planned for 2018 (see, e.g., Molaro 2009). Higher resolution in the data will hence directly influence the positioning errors (see Chapter Analysis II). Commonly, the weighted fits neglects the error in K_i completely, which may become inappropriate in the near future.

Effective analysis involves consideration of the error budget of the sensitivity coefficients. The χ^2 merit function for the generic case of a straight-line fit with errors in both coordinates is given by:

$$\chi^2(a,b) = \sum_{i=0}^{N-1} \frac{(y_i - a - bx_i)^2}{\sigma_{yi}^2 + b^2 \sigma_{xi}^2} \tag{5.3}$$

where σ_{xi} and σ_{yi} are, respectively, the x and y standard deviations for the ith point. The weighted sum of variances in the denominator of Equation 5.3 can be understood as the variance in the direction of the smallest χ^2 between each data point and the with slope b, and also as the variance of the linear combination $y_i - a - bx_i$ of two random variables x_i and y_i,

$$\text{Var}(y_i - a - bx_i) = \text{Var}(y_i) + b^2 \text{Var}(x_i) = \sigma_{yi}^2 + b^2 \sigma_{xi}^2 \equiv 1/w_i \tag{5.4}$$

The sum of the square of N random variables, each normalized by its variance, is thus χ^2-distributed.

Minimizing Equation 5.3 with respect to a and b turns out to be difficult, since the occurrence of b in the denominator makes the resulting equation for the slope $\partial \chi^2 / \partial b = 0$ nonlinear. The corresponding condition for the intercept, $\partial \chi^2 / \partial a = 0$, is still linear and yields:

$$a = \frac{\sum_i w_i (y_i - bx_i)}{\sum_i w_i}, \tag{5.5}$$

where w_i is defined by Equation 5.4. The procedure is now to minimize the in general one-dimensional function to minimize with respect to b, while using Equation 5.5 at each stage to ensure that the minimum with respect to b is also minimized with respect to a.

If any datapoints have very small σ_y but moderate or large σ_x, then it's also possible to have a maximum in χ^2 near zero slope. In that case, there can under certain conditions be two χ^2 minima, one at positive slope and other at negative. Only one of these is the correct global minimum. It is therefor important to have a good starting guess for b. The strategy that is used, is to scale the y_i's so as to have variance equal to the x_i's, then to do a conventional linear fit with weights derived from the scaled sum $\sigma_{yi}^2 + \sigma_{xi}^2$. This yields a good starting guess for b if the data are even plausibly related to a straight-line model.

Finding the standard errors σ_a and σ_b on the parameters a and b is more complicated. They can be expressed as the respective projections onto the a and b axes of the "confidence boundary" where χ^2 takes on a value one greater than its minimum $\Delta\chi^2 = 1$. These projections follow from the Taylor series expansion

$$\Delta\chi^2 \approx \frac{1}{2}\left[\frac{\partial^2\chi^2}{\partial a^2}(\Delta a)^2 + \frac{\partial^2\chi^2}{\partial b^2}(\Delta b)^2\right] + \frac{\partial^2\chi^2}{\partial a\partial b}\Delta a\Delta b. \tag{5.6}$$

The procedure was implemented in C (`2DimErrFit`).

This rather complex approach is not required for the current state of analysis, since other sources of error outweigh the influence of erroneous K_i values by far. At the current level even an error in K_i of about 10% merely has an impact on the error estimate in the order of a few 10^{-6}, as resulted from simulations.

The factual errors are expected to be in the order of merely a few percent (see section 2.2.4), yet they might contribute to the precision of future analysis.

Alternatively to the actual fit with errors in both coordinates, the uncertainties in K_i can be translated into an uncertainty in redshift via a previously fitted slope:

$$\sigma_{z_i\,\text{total}} = \sigma_{z_i} + b \times \sigma_{K_i} \quad \text{with} \quad b = (1 + z_{\text{abs}})\frac{\Delta\mu}{\mu}. \tag{5.7}$$

The results of this ansatz are similar to the fit with errors in both coordinates and in general this is simpler to implement. Due to its simplicity this feature was added to `2DimErrFit`.

Another possibility is to apply a gaussian error to each sensitivity coefficient and redo the normal fit multiple times with alternating variations in K_i. Again, the influence on the error-estimate for a gaussion distributed error of 10% is in the order of 1 ppm.

The different approaches to the fit allow to estimate its overall robustness as well since they all yield identical results.

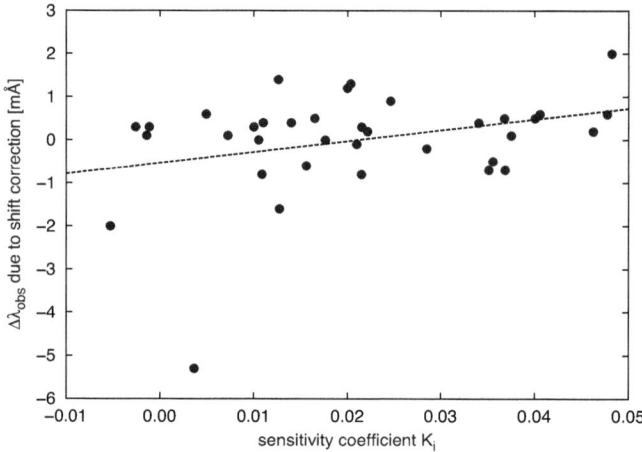

Figure 5.2: Variation in fitted positions for all lines with and without initial correction for shifts in between the 15 spectra. The slope of the exemplary fit is dominated by three lines.

5.4 Influence analysis of data preprocessing

Section 3.2.2 describes the initial shift to a common mean of all 15 spectra. The complete analysis was redone with error-scaled but unshifted spectra and the ascertained line positions of both runs compared. Figure 5.2 shows the difference for each H_2 line in mÅ over the corresponding sensitivity coefficients K_i. The plotted line is a straight fit. Clearly the slope is dominated by three individual lines whose fitted centroids shifted up to 5.5 mÅ due to the preprocessing. These three lines in particular produce a trend towards variation in μ when grating shifts and other effects are not taken into account. This single-sided trend probably occurred by mere chance but at such low statistics it influences the final result. Similar effects might have introduced trends of non-zero variation in former works (see, e.g., Ivanchik et al. 2005; Reinhold et al. 2006).

5.5 Rotational Levels - medium dependent

As for the systematic effects underlying the data, one might envision that the absorbing H_2 cloud is inhomogeneously distributed in cold and somewhat warmer parts, located at slightly differing redshifts. Temperature plays a role through the Boltzmann distribution over populated rotational states in the ground state that take part in the absorption. Here the para-ortho distribution must be taken into account; hence even at the lowest temperatures the $J = 1$ rotational state is significantly populated. Therefore the data set is divided in two separate groups, where the $J = 0$ and $J = 1$ states are referred to as cold, and the $J \geq 2$ states as warm. Figure 5.3 shows the result of the analysis with different symbols for the states referred to as

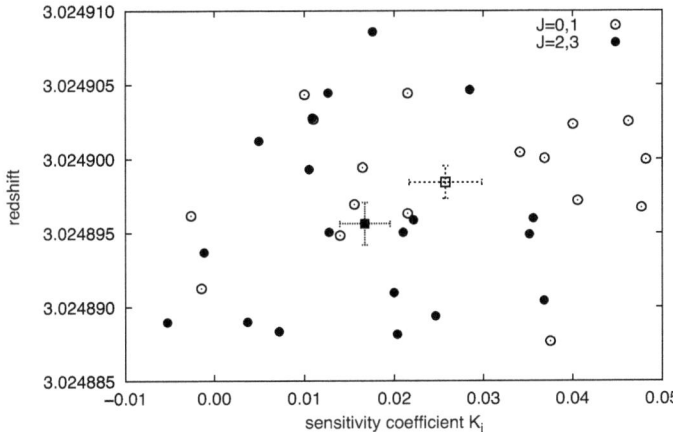

Figure 5.3: Obtained redshift for the rotational states J=0,1 (*open circles*) and the rotational states $J \geq 2$ (*filled circles*) together with the corresponding average values *squares*) for both groups with their standard error.

cold (*open circles*) and the higher rotational states (*filled circles*). The corresponding square data points represent the average in redshift and sensitivity of all lines of the rotational group. These results lead to the conclusion that there is no significant temperature effect underlying the data, since the average redshifts do not differ significantly.

However, taking into account that the average sensitivity of the lines in each group differs, the combination of these two rotational groups does introduce a trend towards positive variation. Not to a truly significant amount but yet mimicking a tendency for a variation in μ. The bootstrap analysis with 50.000 samples of both groups in Figure 5.4 shows that the groups for themselves, analysed separately give a notably lower value for $\Delta \mu / \mu$ with a larger error which can be attributed to the smaller samples.

The individual analysis of the two rotational groups hence is in disagreement with the assumption that an underlying variation in μ causes the different averages of the measured redshifts. Both rotational groups cover a large range in sensitivity ($-0.003 - 0.048$ for $J = 0, 1$ and $-0.005 - 0.037$ for $J \geq 2$) and are hence sensitive to variation in themselves.

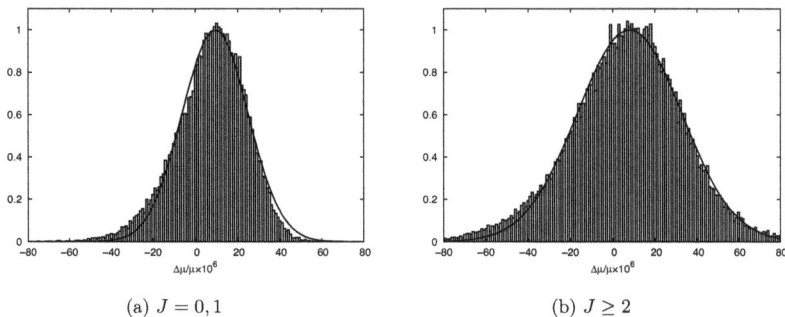

(a) $J = 0, 1$ (b) $J \geq 2$

Figure 5.4: $\Delta\mu/\mu = 9.4 \pm 13.3 \times 10^{-6}$ via bootstrap analysis with 50.000 samples for the 17 lines of rotational level $J = 0, 1$ (*left*) and for the rotational levels $J \geq 2$ (*right*) with 19 lines, the analysis yields $\Delta\mu/\mu = 8.3 \pm 21.1 \times 10^{-6}$

5.6 Vibrational Levels - energy dependent

An important step is to investigate how the reduced redshifts correlate with energy, e.g., photon energy or wavelength. Such an effect may hint at a possible linear calibration error in either one of the spectra.

Alternatively an energy dependence might indicate a process by which the frequency of the light traveling from high-z to Earth-bound telescopes is shifted in a different amount for each frequency component. Such processes are difficult to conceive. Gravitational redshift effects give rise to shifts by $(1+z)$ but these phenomena are independent of frequency. The same

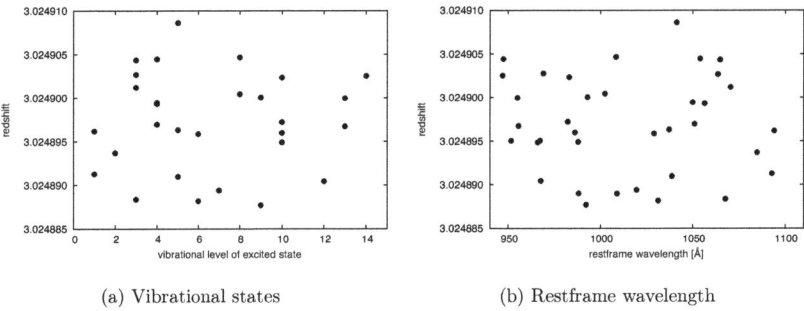

(a) Vibrational states (b) Restframe wavelength

Figure 5.5: Observed redshift vs. the observed vibrational states (*left*), and plotted against the restframe wavelength directly (*right*).

holds for Rayleigh and Raman scattering processes, where the intensity of the scattered light scales with λ^{-4}; the scattering may induce frequency shifts corresponding to rotational and vibrational quanta of the molecules present in the medium, but will not cause any gradual shifts in a spectral distribution. Brillouin scattering is an effect known to produce frequency shifts of the order:

$$\Omega_{\rm B} = \frac{4\pi n}{\lambda} v \sin \frac{\theta}{2} \qquad (5.8)$$

with n the index of refraction, λ the wavelength of the propagating light beam, v the speed of sound, and θ the scattering angle (see Boyd and Braun 1992). This could produce a maximum frequency dependent shift of about 1 cm^{-1}, for a typical speed of sound for a gaseous medium of $v = 300$ ms^{-1} and a wavelength of $\lambda = 1000$ Å, an amount that could well disturb the analysis of the quasar data. However, in the forward scattering direction ($\theta = 0$), in which the quasar light is detected on Earth, the Brillouin shift is zero. Moreover, the density along the line of sight is very low, so that it is difficult to conceive how Brillouin intensity could be produced. If Brillouin scattering were to be a dominant process, the light would be scattered away over a 4π solid angle. Under the required conditions of high density, absorption would dominate at wavelengths in the extreme ultraviolet range, so that no light could be detected from the quasar systems.

Nevertheless a correlation analysis of the apparent redshifts z versus photon energy was performed for this data. Figure 5.5a plots the measured redshifts of the Lyman transitions versus their corresponding vibrational level of the excited state, whereas Figure 5.5b graphs the redshifts against the laboratory transition energies directly. Evidently, there is no correlation between the apparent redshifts and the photon energy.

5.7 Electronic levels - Lyman, Werner bands

Figure 5.6 plots the result of the analysis indicating the observed Lyman (*filled circles*) and Werner (*open circles*) lines. Additionally their average in redshift and sensitivity towards $\Delta\mu$ is plotted (*squares*).

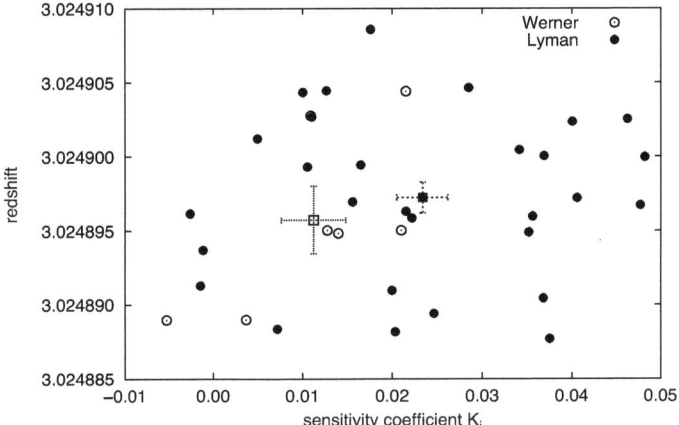

Figure 5.6: Obtained redshift for the Lyman lines (*filled circles*) and the Werner band (*open circles*). The *squares* represent the corresponding average values with their standard error.

The average redshifts of the different electronic transitions do not differ significantly. Together with their offset in the covered sensitivity region their discrepancy introduces a trend towards positive variation though.

That is the expected behavior for the case of a true underlying variation in μ, but an important issue is that the mere Lyman band shows no indication of variation (see Fig. 5.7) even though it covers the whole range in sensitivity space from $K_i = -0.003$ to $K_i = 0.048$ (see Fig. 5.6). The impression of a large scatter in the measured redshifts of the Werner lines can be confirmed with the bootstrap analysis. Figure 5.7 is based on a subset of 29 lines, excluding the 7 Werner band lines. The best fit for $\Delta\mu/\mu$ is notably smaller while the width of the distribution remained about the same despite the lower number of lines taken into account. The corresponding best fit gives: $\Delta\mu/\mu = (5.4 \pm 12) \times 10^{-6}$.

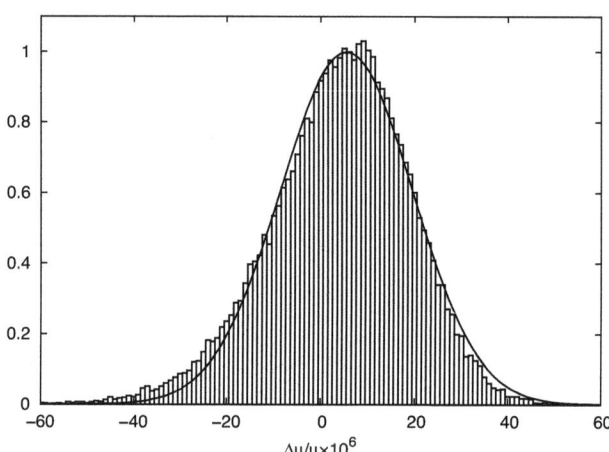

Figure 5.7: $\Delta\mu/\mu$ and in particular its uncertainty derived via 10.000 bootstrap samples of the Lyman lines only. The gaussian fit yields a centroid at 5.4 and a σ of 11.99.

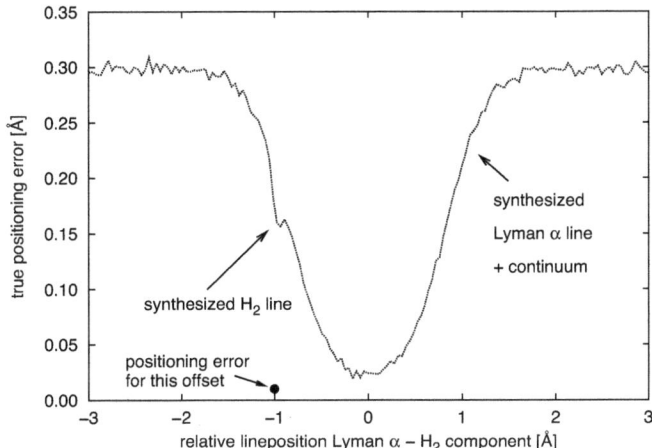

Figure 5.8: The graph (*dotted line*) shows an exemplary case of the 10.000 runs with the H$_2$ component at an offset of -1 Å with regard to the Lyman-α component. The *filled circle* indicates the resulting positioning error of a fit on the left axis.

5.8 Accuracy of line fits

To test the accuracy of fits without a clean continuum a code to generate synthetic spectra was written and several simulations were performed (`SpecSim`).

For that purpose a synthesized molecular hydrogen line was placed in the region of the wing of a broad almost saturated synthesized Lyman-α feature as illustrated in Figure 5.8. To meet the conditions of the combined observed quasar spectra, a signal-to-noise of about 90 containing Gaussian and Poisson photon noise were synthesized. The intention of this simulation is to reveal systematic effects rather than statistical errors. The position of the synthesized H$_2$ feature was shifted through the broad Lyman-α line in 100 steps of 0.05Å. For each position a number of 100 synthesized spectra were generated and each time the molecular hydrogen line along with the atomic component was fitted. For each fit the absolute deviation of the fitted H$_2$ position and its true position as was fed into generation of the synthetic spectra is plotted against the true position. This allows to determine the accuracy and stability of the fitting algorithm and its dependency on the line environment. In the ideal case of a known and clean atomic hydrogen component it is possible to fit the parameters of the DLA component coevally. This procedure, however, is not available for most DLA systems that show H$_2$ absorption due to heavy contamination of the continuum.

Figure 5.9 shows the resulting positioning errors of the simulations. On the left the estimated errors in position for the two-component fit are plotted in comparison with the true offset. The mean true error of 100 simulations per set up can be easily derived by the known line position as incorporated into the synthesized spectrum and the fitted parameter for the center of the line. As can be seen the fits with the H_2 component placed in a clean continuum result in an

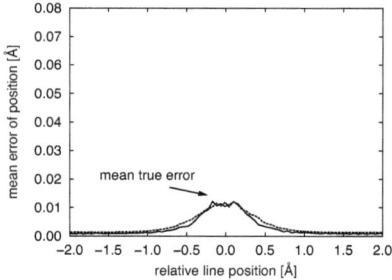
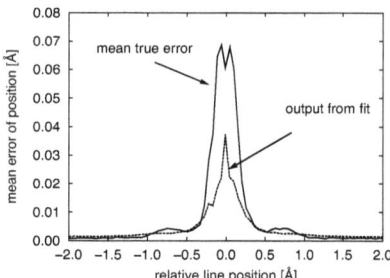

Figure 5.9: Mean value of true error and outcome of the fitting procedure. Comparison of two component fit (*left*) and single component fit (*right*).

constant average error of a few mÅ. However, the amplitude of the error distribution increases towards the center of the central component. As the simulation shows, up to a certain point the fitting accuracy of the H_2 line is not affected by its position on the outer wing of the broad Lyman-α line.

For Figure 5.9 (*right*) the same method was used to generate the spectra but instead of fitting both components and the continuum simultaneously, the continuum is fitted by a third grade polynomial to the flux enclosing the H_2 line without fitting the Lyman-α component explicitly. The fits carried out in this thesis are based on single-component fits, meaning that for each H_2 absorption feature, parameters for a single H_2 line are fitted.

This is a dramatic difference since for that approach no information on the environment of the fitted line is needed. The physical origin of the continuum contamination is not taken into account here. As Figure 5.9 shows, the quality of the fits obtained via polynomial fit (*right*) and the two component fit (*left*) is quite similar in the clean continuum area and up to a certain point on the wing of the broad Lyman-α feature. This is a beneficial result since in the observed spectra it would be impossible to divide the continuum contamination into single components that could be fitted simultaneously. However, the position of the H_2 lines is critical for the quality of the fit and they cannot easily be determined. Figure 5.9 also reveals the different behavior towards the center of the main component. The comparatively low equivalent width of the H_2 line makes it practically indistinguishable from noise near the core for the one component fit. In the core area the optically thin H_2 feature is practically unobservable. The method of polynomial continuum fit is clearly inferior near the core of the main component in this simulation

The accuracy of a fitted line position depends on the continuum contamination or rather its position on a wing of a stronger, in DLA-systems saturated Lyman-α component (see, e.g,

Wendt and Reimers 2008). Near the center of such a Lyman-α feature, the fitting method has a great influence on the error distribution.

The resultant influence on the fit is reviewed on the basis of the generated synthetic spectra. Figure 5.9 already showed the mean true error of the simulation runs as gained from exact comparison (*bold*) and the mean value of the calculated error from the output of the fitting program (*dotted*).

The procedure of fitting a single component only and describing the continuum by rectifying a third grade polynomial fit leads to significantly larger errors in the core area. Also the mean error of line position fits for lines situated near the core region is dramatically underestimated by the fitting program. In the case of the two component fit the mean true error and the error given by the fitting routine match quite well. Great caution was exercised in the selection of H_2 lines to avoid such critical cases. The approach presented here, avoids such problematic cases since the positioning near saturated areas is in general not very stable however, and these lines failed the 3-σ-criterion entirely. It is evidently difficult to evaluate the relative position of a H_2 feature, but in Ivanchik et al. (2005) and Thompson et al. (2009a) such potentially problematic lines are excluded manually as well. In Figure 5.10 the errors of the simulated

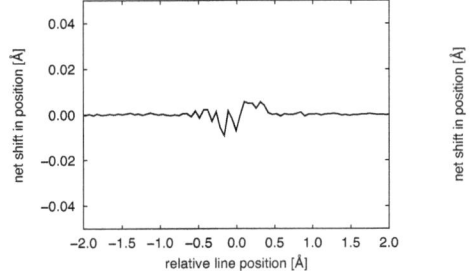
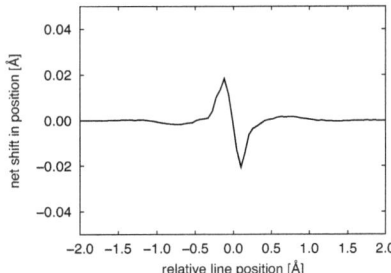

Figure 5.10: Mean shifts of fitted position over a series of 10.000 fits for the two component fit (*left*) and the single component fit (*right*).

fits were added up to reveal a possible net shift of the position fit. The results for the two methods differ notably. While the two component fit on the left shows no significant overall shift despite the greater uncertainties in the core region, the one component fit produces a net shift of the fitted line position. A potential shift in the order of the average error in clean continuum demonstrates the importance of avoiding lines in such critical positions.

The apparent solution to use a multi component fit and thus to 'recreate' the continuum level is only applicable in few cases. The ideal conditions of a single Lyman-α as simulated do not correspond to the observed data. Fitting an arbitrary number of free Lyman-α components in the environment of each H_2 line could also easily result in an shift of the comparatively weak H_2 lines and does not represent the physical conditions of the absorber. The uncontrolled fit to an alternating number of parameters for a complex region of absorption, confines the fitting procedure to some sort of a black box and may produce unjustified confidence in the outcome. The only acceptable consequence is to avoid these critical lines, since even for the simulated two component fit the individual error in position is unacceptable for the aspired precision.

At these low statistical totalities systematicly induced shifts are unlikely to even out. However, they would not mimic a variation of μ since this effect applies equally to all transitions. In general the simulations show that the mean error as estimated by the fitting procedure seems to be rather reasonable for unblended, non-critical lines.

6 Analysis II

6.1 Data

6.1.1 2009 observations

The recent observations of QSO 0347-383 have been performed with UVES on VLT on the nights of September 20-24 2009[1]. The journal of these observations as well as additional information on the detected relative shifts (see section 6.2.1) is given in Table 6.1. The DIC 2 setting was used with blue setting at 437 nm grating angle. The images preserved the original pixel size. of $0.013 - 0.015$ Å, pixel, or 1.12 kms^{-1} at 400 nm along dispersion direction. The UVES observations comprised of 10×5400 second-exposures on four successive nights and 1 exposure of 3812 seconds. Eight UVES spectra were taken with DIC2 and setting 437+760 and three with the 437+860, thus providing blue spectral ranges between 373-500 nm. QSO 0347-383 has no flux below 370 nm due to the Lyman discontinuity of the $z_{abs} = 3.023$ absorption system. The slit width was set to 0.7″ for all observations providing a Resolving Power of $\sim 65554 \pm 3868$

Table 6.1: Journal of the observations (2009 data). Before and after each spectrum, a 30 sec calibration frame was recorded

No.	Date	Time	λ	Exp[sec]	shift to °1 [mÅ]
1	2009-09-20	05:05:46	437	5400	0.0
2	2009-09-20	08:28:48	437	3812	0.1
3	2009-09-21	04:45:51	437	5400	-0.9
4	2009-09-21	06:18:45	437	5400	1.1
5	2009-09-21	07:59:24	437	5400	8.3
6	2009-09-22	04:41:37	437	5400	-2.5
7	2009-09-22	06:14:25	437	5400	-1.4
8	2009-09-22	07:59:19	437	5400	-5.8
9	2009-09-23	04:24:05	437	5400	-6.9
10	2009-09-23	05:56:49	437	5400	1.9
11	2009-09-23	07:29:35	437	5400	6.0

in the blue frame. The seeing was varying in the range between 0.5″ to 1″ as measured by

---

[1] Program ID 083.A

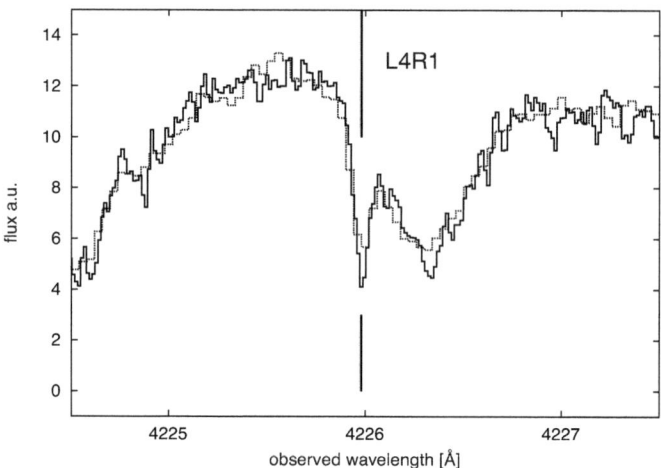

Figure 6.1: The region around the observed H_2 absorption feature corresponding to the L4R1 transition plotted in the co-added data of 2009 (*solid*) and the same region in the 2002 data set (*dots*).

DIMM but it normally is better at the telescope.

Figure 6.1 demonstrates increase in resolution in comparison to the 2002 data of section 3.1.1. The 11 different spectra and the corresponding co-added data (*bottom*) are shown in a region around L4R1 in Figure 6.2.

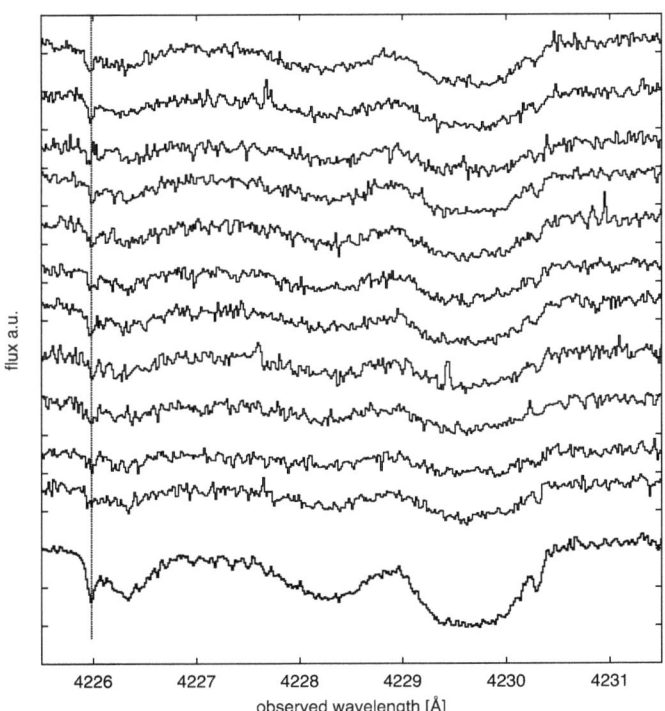

Figure 6.2: The 11 single spectra and the corresponding co-added data (*bottom*) are plotted around the region of L4R1 (*vertical line*). H_2 features cannot be distinguished in single spectra.

6.1.2 Reduction

The last version of the UVES pipeline has been followed for the data reduction. The pipeline first uses a set of 5 available bias to make a master bias that is free of cosmic ray hits which is then subtracted to all frames. In our case the two dimensional format images of the long-slit calibration lamp images and long-slit flat-field images have been bias subtracted but not flat fielded. The pipeline utilize a pinhole lamp for identifying the location of the orders. The order traces are along the x direction but are curved and tilted upward.

Often the ThAr lamps exposures are taken during daytime, which means several hours before the science exposures and likely under different thermal and pressure conditions. The paths for ThAr light and quasar light through the spectrograph are not identical thus introducing small distortions between ThAr and quasar wavelength scales.

See section 3.1.2 for more details on the residuals.

Calibration distortions have been investigated at the Keck/HIRES spectrograph by comparing the ThAr wavelength scale with one established from I2-cell observations of a bright quasar by Griest et al. (2010). In the wavelength range $\sim$ 5000 − 6200 Å covered by the iodine cell absorption they found both absolute offsets which can be as large as 500 − 1000 m s^{-1} and an additional saw-tooth distortion pattern with an amplitude of about 300 m s^{-1}. The distortions are such that transitions at the order edges appear at different velocities with respect to transitions at the order centers when calibrated with a ThAr exposure. Whitmore et al. (2010) recently repeated the same test for UVES with similar finding though the saw-tooth distortions show slightly reduced peak-to-peak velocity variations of $\sim$ 200 m s^{-1}. The physical explanation for those distortions is not yet known, so it remains to be examined whether the deviations are the same at other wavelengths or depend for the specific exposure. Indication that these offsets do not appear to apply for the observations at hand can be found in section 8.4.

Examination of the UVES spectrograph at the VLT carried out via solar spectra reflected on asteroids with known radial velocity showed no such dramatic offsets, being less than $\sim$ 100 m s^{-1} (Molaro et al. 2008b) but systematic errors at the level of few hundred m s^{-1} have been revealed also in the UVES data by comparison of relative shifts of lines with comparable response to changes of fundamental constants (Centurión et al. 2009).

If we assume that similar intra-order distortions apply to our spectra of at much bluer wavelengths then, because the molecular transitions of interest lie at different positions along different echelle orders, we should expect the effect to increase the line position scatter of the lines around the mean redshift. A peak-to-peak intra-order value of 200 m s^{-1} or a σ of 70 m s^{-1} corresponds to an additional error in $\Delta\mu/\mu$ of approximately $\pm 4.7 \times 10^{-6}$.

The goal of the wavelength calibration is a proper vacuum wavelength at rest with respect to the barycenter of the earth sun system. For the data presented in this thesis, Paolo Molaro carried out the extended procedure. First the pixel-wavelength conversion is done by using the associated long slit calibration spectrum. Once determined from the ThAr exposures, the wavelength solutions are simply applied to the corresponding quasar exposures. Murphy et al. (2008) and Thompson et al. (2009a) have independently shown that the standard Th/Ar line list used in the old UVES pipeline analysis was a primary limiting factor in obtaining the accuracy required for a determination of μ at the 10^{-5} level. Thompson et al. (2009b) recalibrated the wavelength solutions using the long slit calibration line spectra taken during the observations of the two QSOs and argued that the new wavelength calibration was the primary reason for a null result in their study. The new data UVES pipeline has solved some of these problems and has been adopted in the calibrations in place.

The blue frame comprise 32 orders from absolute number 96 to 124 covering the wavelength range 374-497 nm while the molecular lines are spread over only 18 UVES echelle orders (106-122) covering the wavelength range 380-440 nm. In this One cannot assume possible shift caused by this to even out completely range about 395 ThAr lines, more than the 55% of the lines in the region were used to calibrate the lamp exposures. A polynomium of the 5th order was adopted and typical residuals of the wavelength calibrations were of ~ 0.34 mÅ or ~ 24 m s^{-1} at 400 nm and are consistent with being symmetrically distributed around the ïñĄnal wavelength solution at all wavelengths. This is by far the most precise wavelength calibration obtained. By comparison in Malec et al. (2010) the wavelength calibration residuals are RMS ~ 80 m s^{-1}.

In our set of observations calibration spectra were taken before and after the object spectra for each night. Observing the calibration lamp immediately before and immediately after an object observation provides an accurate monitor of any time variation in the wavelength calibration. Moreover, the calibration frames were taken in the attach mode avoiding spectrograph resetting at the start of every exposure. Since Dec 2001 UVES has implemented an automatic resetting of the Cross Disperser encoder positions at the start of each exposure 4 . This implementation has been done to have the possibility to use daytime ThAr calibration frames for saving night time. If this is excellent for standard observations, it is not for the measurement of fundamental constants which require the best possible wavelength calibration. Thermal-pressure drifts move in different ways. The different cross dispersers thus introduce relative shifts between the different spectral ranges for different exposures. Only frames observed in a sequence avoid automatic resets of grating positions. The reported encoder readings indicate that there were no grating resets within each block of observations.

It should be emphasized that this effect has not been taken into account for all the analysis performed so far on UVES data either for measuring α or μ variability.

Calibration frames taken immediately before and after the science frame minimize the inïñĆuence of changing ambient weather conditions which cause different velocity offsets. In this way it is possible to track the shifts in the wavelength positions between observations and between different observing nights. There are no measurable temperature changes for the short exposures of the calibration lamps but during the much longer science exposures the temperature drifts generally by 0.1 K, and in two cases the drift is of 0.2 K and in other two there is no measurable change. Pressure values are surveyed at the beginning and end of the exposures and changes range from 0.2 to 0.8 mbar. The estimates for UVES are of 50 m s^{-1} for $\Delta T = 0.3$ K or a $\Delta P = 1$ mbar (see UVES manual, Kaufer et al. 2004), thus assuring a radial velocity stability within ~ 50 m s^{-1}.

Individual spectra are corrected for the motion of the observatory about the barycenter of the Earth-Sun system and then reduced to vacuum. The component along the direction to the object of the barycentric velocity of the observatory was calculated using the date and time of the midpoint of the integration. This velocity is due to the earthâĂŹs orbit and rotation relative to the barycenter of the earth-sun system. The wavelength scale was then corrected for this motion so that the ïňĄnal wavelengths are vacuum wavelengths as observed in a reference frame at rest relative to the barycenter. The air wavelengths have been transformed to vacuum by means of the dispersion formula by Edlén (1966). Drifts in the refractive index of air inside the spectrograph between the ThAr and quasar exposures will therefore cause miscalibrations. The temperature and atmospheric pressure drifts during our observations were < 1 K and < 1 mbar respectively. According to the Edlén (1966) formula for the refractive index of air, this would cause diïňĄerential velocity shifts between 370 nm and 440 nm of ~ 10 m s^{-1}, which are negligible for today's precision.

These corrections require a rebinning which introduces a certain degree of correlations between the flux of neighbouring pixels. The effect on the fitted centroid of spectral features has been tested with `SpecSim` and Malec et al. (2010) and found to be small, namely lower than $\pm 0.8 \times 10^{-6}$.

These requirements of high wavelength accuracy sets strong demands on the conditions of the observation in general. Lines on the CCD od the spectrograph are images of the slit and in UVES a 1 arcsec slit has a FWHM of about 7.5 km s^{-1}. Thus a shift in the centroid of image position of 10% already provides a shift of 750 m s^{-1}.

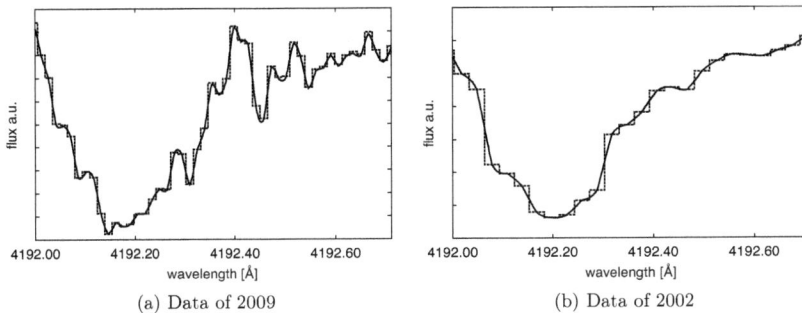

Figure 6.3: The original flux is interpolated by a polynomial using *Neville's algorithm* to conserve the local flux (see section 3.2.2). Panel a) shows an exemplary region in a single spectrum of the recent data, panel b) corresponds to the identical region in one of the 15 spectra of 2002.

6.2 Preprocessing of data

6.2.1 Correction for individual shifts

The data recorded in 2009 was checked for relative shifts analogous to section 3.2.2. Again, each spectrum was subsampled via *Neville's algorithm* and the shift with the best conformity was ascertained (ShiftCheck). Figure 6.3 displays a sample region of the new data (6.3a) and the data described in section 3 (6.3b). The resulting corrections comprise an average deviation of about 0.232 kms^{-1} and are listed in Table 6.2.

6.2.2 Selection of H$_2$ lines

The selection of suitable H$_2$ features for the final analysis is rather subjective. Figure 6.5 plots a region with 3 H$_2$ components in the dataset of 2002 and in the 2009 observations. The doubling in the effective resolution reveals an potential blend with another line in the absorption wings in the case of L5P1, which might have caused this particular line to be excluded by the robustness-of-fit criterion in section 3.2.3.

For the analysis the line selection based on visual inspection of the data. At the individual signal-to-noise ratio of about 8 the H$_2$ features are hardly detectable in the single spectra. Figure 6.2 illustrates that. For the purpose of line selection and identification the 11 spectra were rebinned and co-added (CoAdd) after the procedure described in the previous section was applied.

To identify the observed H$_2$ transitions and distinguish them from the Lyman-α forest a huge database format for all eligible H$_2$ lines was created. It was based on tables of more than

Table 6.2: Relative shifts of the observed spectra in 2009 to their common mean.

Spectrum	shift to mean [kms^{-1}]
1	-0.0101
2	-0.0032
3	-0.0730
4	0.0699
5	0.5963
6	-0.1943
7	-0.1129
8	-0.4358
9	-0.5174
10	0.1315
11	0.4279
average deviation	0.2324

1.000 transitions that Abgrall et al. (2000) made available in electronic form at the CDS. The data were converted into wavelength and the corresponding latest sensitivity coefficients by Varshalovich and Levshakov (1993); Reinhold et al. (2006); Meshkov et al. (2006) and Ubachs et al. (2007) were incorporated. The rest frame wavelength were updated via the data presented in Hollenstein et al. (2006); Ubachs et al. (2007) and very recently Bailly et al. (2010). Today, this database includes all transition wavelengths including their oscillation strengths up to the 17th vibrational level and the 15th rotational quantum number in Lyman and Werner bands. By now, sensitivity coefficients for all observed H$_2$ lines in quasar absorption spectra are included as well. To locate all possible H$_2$ features in a data set a graphical data examination tool (GRADE) was written that displays all lines up to a given vibro-rotational state with a preassigned redshift along with the co-added data. Figure 6.4 is a monochrome screenshot of that program. It shows the region centered at line L4R1. The expected observed position is shown as a vertical line and in numbers for L4P1 and l4R2. This example illustrates the small differences for rotational energies as well. The energies for electronic, vibrational and rotational transitions differ roughly by a factor of 1.000 to smaller values. In the plotted case 3 lines from 4th vibrational state in that region are selected and the corresponding sensitivity coefficients are shown in the *top*. L4R0 is not suited for μ analysis at all for its location on a wing of Ly-α absorption and L5R5 cannot be detected.

The *lower left* of Figure 6.4 shows additional info of the selected line, such as its listed laboratory wavelength, the oscillator strength as well as the deviation of a previous fit from the expected position in mÅ and the corresponding fitting error. For this purpose the resulting table of a fit can be read as reference. The selected lines can be arranged in groups to differentiate between different selections or to fit test groups and the region of appendant continuum along with the

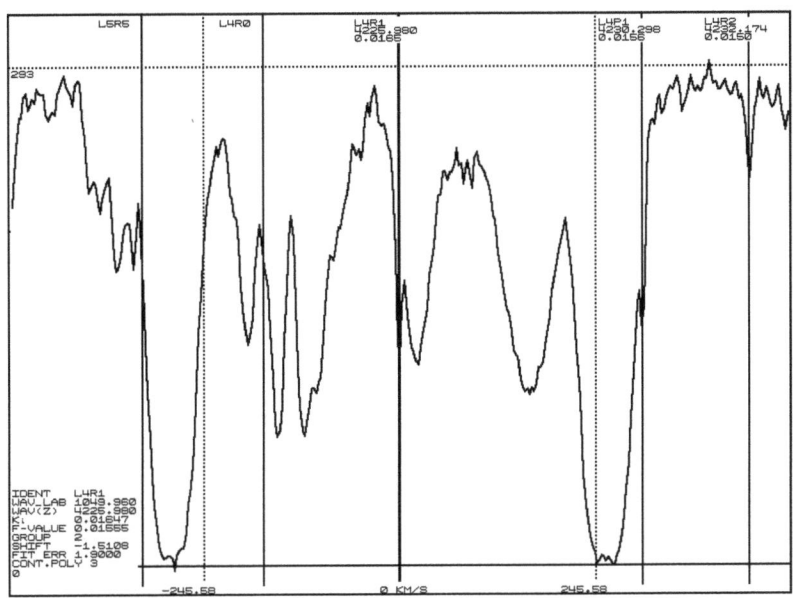

Figure 6.4: GRAaphical Data Examination tool for line selection (GRADE).

suited grade for a polynomial fit can be defined for each line separately. The line data is read from the aforementioned own database and the selection can be exported into tables suitable to generate input data for RQFit.

The information on the lines selected for fitting and further details about continuum modeling are used to prepare detailed scripts for the final fit (ScriptGen). Lines of the same rotational level share one fit parameter for the broadening width and the column density along the line of sight for all simultaneously fitted spectra. The observed redshift is of course handled as free parameter per line, but fixed for the separate spectra. The high resolution data allows for a comparison with former multi-component fits carried out by King et al. (2008). They fit numerous additional components in a region of H_2 absorption to narrow down the χ^2 of the fit to the data. In fact the evolution of χ^2 with an increasing number of additional free lines is their only criterion to fix the total number of components. Some of the resulting issues are described in section 3.3.2 on page 38. A higher resolution may verify or falsify some of the decisions on additional components and help to distinguish between apparent precision (lower χ^2) and reached accuracy (better description of the physical conditions of the absorber).

Figure 6.6 plots the data mentioned in section 3.1.1 and used in the analyses by Ivanchik et al. (2005); King et al. (2008) and King et al. (2008). The *solid* vertical line marks the H_2 component L4R1 and the *dotted* lines indicate the 13 additional components in that region. The *upper* plot corresponds to the data of 2009 and reveals that some of the extra components

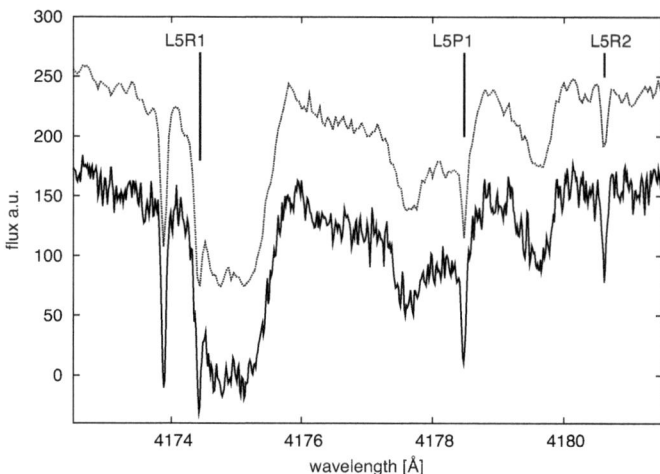

Figure 6.5: Plot of the 15 co-added and rebinned spectra of 2002 (see Fig. 3.8 and, with an offset, the recent data of 2009 (*solid*), composed of 11 rebinned and co-added spectra for illustration.

evidently recreate the flux observed in 2002 but do not correspond to factual properties of the absorber. The parameters of the additional components as well as the obtained redshifts of the H_2 components are undisclosed and hence cannot be reviewed further. The positions of the addition components were extracted from a plot in King et al. (2008) at pixel-accuracy.

6.2.3 Resolution

The new observations in 2009 and its careful reduction in 2009/10 provides the best data available for QSO 0347-383. Apparent features revealed by the higher resolution had to be tested for significance since the signal-to-noise ratio is comparably low as a trade-off to 1×1 binning and its associated gain in resolution. A region near 4227 Å qualifies as exemplary case. Figure 6.7 shows the corresponding co-added spectrum at *top*. The vertical *dashed* lines border the region of particular interest. A couple of absorption signatures could likely be attributed to noise. To verify this assumption, the flux of that region was fitted with a linear function which provided an average flux F_{avg} per pixel. The same was done for the estimated error of flux for the given wavelength interval yielding $\bar{\sigma}_F$. The significance of Flux per pixel i is then given by:

$$\frac{F_i - F_{avg}}{\bar{\sigma}_F}. \tag{6.1}$$

The resulting value per pixel for that region is plotted in Figure 6.7 (*bottom*). The *dotted* lines above and below the zero level mark the 3σ region of significance. In the graphed example the

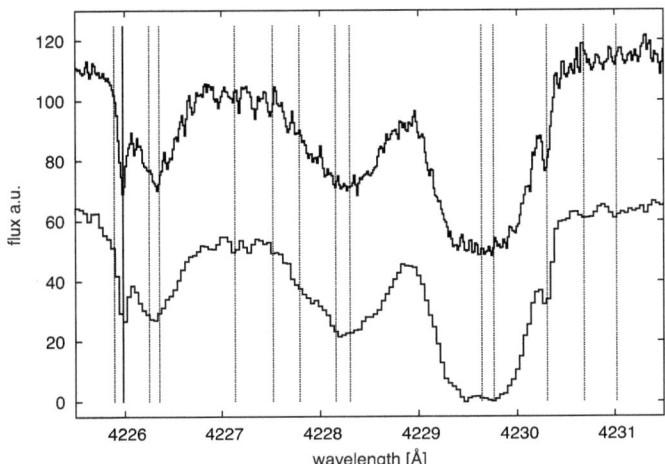

Figure 6.6: Comparison between the 2009 data (*top graph*) and, with an offset, the original single observation run data of 2002 (9 frames, *bottom*) as used by Ivanchik et al. (2005); King et al. (2008); Thompson et al. (2009a). The vertical lines indicate the positions of the H$_2$ component (*solid*) and the 13 additional components fitted in King et al. (2008).

occurring fluctuations in the flux bear no distinct significance. Analogous steps were carried out for several parts of the data (`SigTest`).

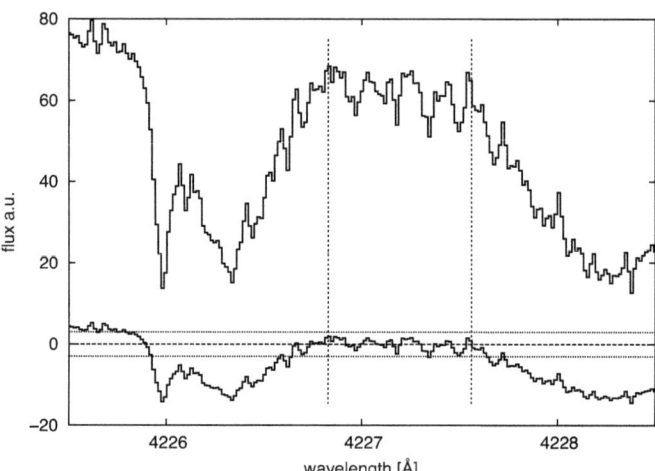

Figure 6.7: Significance test for absorption features in the co-added high resolution spectrum of QSO 0347-383. The *top* plot represents the observed flux with the region of interest marked between the vertical lines. The *bottom* part reflects the difference to the average flux of that region divided by its average error. The *dotted lines* above and below the zero level mark the 3σ divergence.

7 Results II

7.1 Determination of $\Delta\mu/\mu$

Figure 7.1 shows the resulting plot of the measured redshifts of the selected 37 lines with their corresponding sensitivity coefficients K_i. The unweighted fit yields:

$$\Delta\mu/\mu = (7.3 \pm 9) \times 10^{-6}. \tag{7.1}$$

Taking into account the individual positioning error as estimated by `RQFit`, the weighted fit gives:

$$\Delta\mu/\mu = (2.8 \pm 8.3) \times 10^{-6}. \tag{7.2}$$

The analysis with regard to the estimated uncertainties in the sensitivity coefficients (see section 5.3 on page 52) shows no noteworthy deviation from the results in Eq. 7.2 as is expected for the current ratio of positioning errors to ΔK_i.

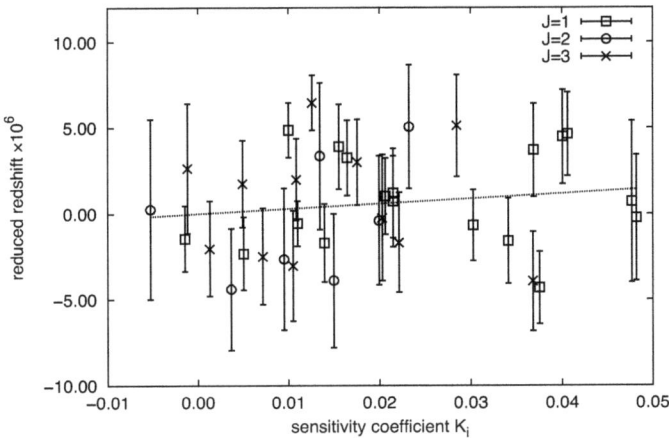

Figure 7.1: The applied unweighted fit to the QSO 0347-383 data corresponds to $\Delta\mu/\mu = (7.3 \pm 9) \times 10^{-6}$. The error bars reflect the positioning error as estimated by the fitting procedure.

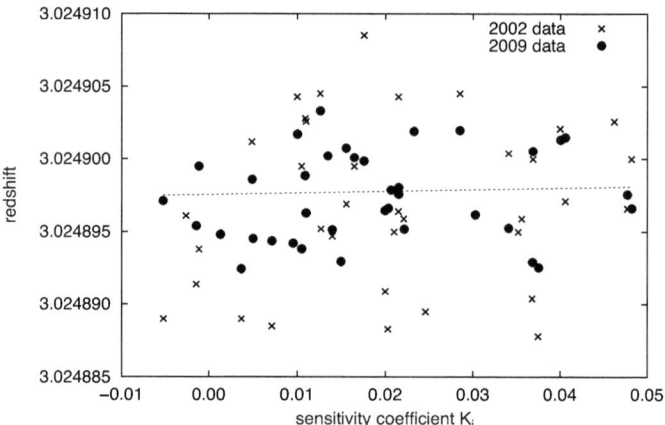

Figure 7.2: Final results in redshift vs. sensitivity coefficient K_i for the analysis of the 2009 data (*filled circles*) in comparison with the 2002 data (*crosses*). The dotted line represents the weighted fit of the 2009 data (see Eq. 7.2).

Figure 7.2 shows the presented results in comparison with those from chapter 4.1. The scatter of the measured redshifts is significantly lower for the recent data (*solid circles*). The gain quality can be attributed to the higher resolution, which implicated a lower signal-to-noise ratio though, and to the greater care with which the observations were carried out as to their specific use and the detailed reduction by Paolo Molaro. The impact of these conditions is examined in the following. The bootstrap analysis of the linear fit to the redshifts is plotted in Figure 7.3. The gaussian fit attests a perfect gaussian distribution of the bootstrap samples and yields:

$$\Delta\mu/\mu = (7.4 \pm 7.3) \times 10^{-6}, \tag{7.3}$$

and is in good agreement with the unweighted fit and confirms its error estimate based on the actual scatter in the data now. The plot also shows the corresponding gaussian fit of the 2002 data (see Figure 5.1 on page 50). The current data evidently provides a more stringent null result.

Table 7.1: QSO 0347-383 Line List (2009 data)

Line ID	K_i	$\lambda_{\rm obs}$ [Å]	$\sigma_{\lambda_{\rm obs}}$ [Å]	$\lambda_{\rm lab}$ [Å]	$\sigma_{\lambda_{\rm obs}}$ [kms^{-1}]	$z_{\rm abs}$
W3Q1	0.0215	3813.2761	0.0025	947.4219	0.194	3.0248976
L13R1	0.0482	3844.0411	0.0035	955.0658	0.274	3.0248966
L13P1	0.0477	3846.6280	0.0045	955.7083	0.352	3.0248976
W2Q1	0.0140	3888.4355	0.0022	966.0961	0.166	3.0248951
L12R3	0.0368	3894.7963	0.0028	967.6770	0.217	3.0248929
W2Q3	0.0109	3900.3250	0.0023	969.0492	0.174	3.0248988
L10R1	0.0406	3952.7519	0.0024	982.0742	0.185	3.0249015
L10P1	0.0400	3955.8151	0.0027	982.8353	0.202	3.0249013
W1Q2	0.0037	3976.4911	0.0035	987.9745	0.262	3.0248924
L9R1	0.0375	3992.7594	0.0021	992.0164	0.158	3.0248925
L9P1	0.0369	3995.9599	0.0027	992.8096	0.205	3.0249005
L8R1	0.0341	4034.7647	0.0025	1002.4521	0.187	3.0248953
L8P3	0.0285	4058.6548	0.0030	1008.3860	0.225	3.0249020
W0R2	-0.0052	4061.2214	0.0053	1009.0249	0.390	3.0248971
L7R1	0.0303	4078.9787	0.0021	1013.4370	0.152	3.0248962
L6P2	0.0232	4138.0250	0.0037	1028.1058	0.266	3.0249019
L6R3	0.0221	4141.5632	0.0030	1028.9866	0.216	3.0248952
L6P3	0.0203	4150.4436	0.0038	1031.1926	0.271	3.0248966
L5R1	0.0215	4174.4222	0.0027	1037.1498	0.193	3.0248980
L5P1	0.0206	4178.4763	0.0023	1038.1571	0.162	3.0248979
L5R2	0.0200	4180.6209	0.0039	1038.6903	0.280	3.0248964
L5R3	0.0176	4190.5599	0.0026	1041.1588	0.187	3.0248999
L4R1	0.0165	4225.9829	0.0023	1049.9597	0.161	3.0249001
L4P1	0.0156	4230.3015	0.0026	1051.0325	0.181	3.0249008
L4R2	0.0150	4232.1689	0.0041	1051.4985	0.290	3.0248929
L4P2	0.0135	4239.3642	0.0045	1053.2843	0.320	3.0249002
L4R3	0.0126	4242.1519	0.0017	1053.9761	0.123	3.0249033
L4P3	0.0105	4252.1852	0.0034	1056.4714	0.241	3.0248938
L3R1	0.0110	4280.3166	0.0014	1063.4601	0.099	3.0248963
L3P1	0.0100	4284.9321	0.0017	1064.6054	0.122	3.0249017
L3R2	0.0095	4286.4914	0.0044	1064.9948	0.310	3.0248942
L3R3	0.0072	4296.4886	0.0030	1067.4786	0.208	3.0248944
L3P3	0.0049	4307.2086	0.0027	1070.1409	0.190	3.0248986
L2R1	0.0050	4337.6244	0.0023	1077.6989	0.161	3.0248945
L2R3	0.0013	4353.7742	0.0030	1081.7113	0.205	3.0248948
L2P3	-0.0011	4365.2462	0.0041	1084.5603	0.282	3.0248995
L1R1	-0.0014	4398.1336	0.0021	1092.7324	0.146	3.0248954

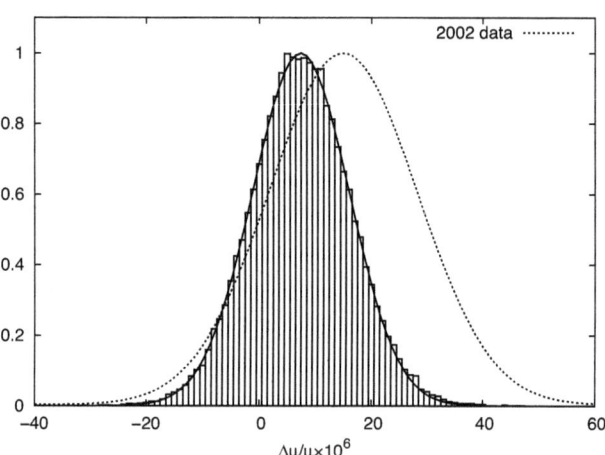

Figure 7.3: $\Delta\mu/\mu$ and in particular its uncertainty derived via 50.000 bootstrap samples. The gaussian fit yields a centroid at 7.4 and a FWHM of 17.2. The *dotted* gaussian corresponds to the accordant bootstrap analysis of the 2002 data (see Fig. 5.1).

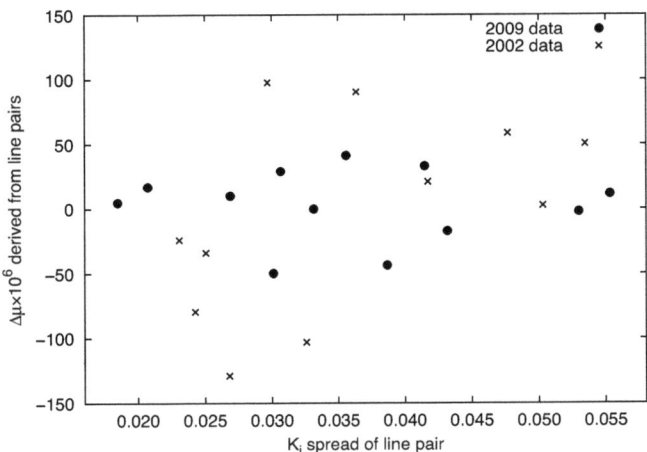

Figure 7.4: The *filled circles* show $\Delta\mu/\mu$ derived from individual line pairs (12) which were selected to give the largest difference in sensitivity ($\Delta K_i \geq 0.018$) towards variation in μ (See Table 7.2). The *small crosses* plot the measurements of 2002. (see Fig. 4.3 in section 4.2).

7.2 Result via discrete line pairs

To obtain a robust value for $\Delta\mu/\mu$, all observed lines are re-sorted by their intrinsic sensitivity towards changes in μ and grouped up in pairs that show the largest separation in K_i (`ToPair`). $\Delta\mu/\mu$ can then be derived from the gradient of Δz_{abs} to ΔK_i for each pair individually. The selected pairs are the most sensitive to variation and are listed in Table 7.2. The pairs were selected regardless of their separation in the observed spectrum which is a crucial point. Their spread in wavelength is listed in Table 7.2 as well.

The resulting measurements of the 12 formed pairs are plotted in Figure 7.4 (*circles*), along with the data of the corresponding analysis in section 4.2 (*crosses*). The 12 line pairs yield via direct fit:

$$\Delta\mu/\mu = (2.8 \pm 8.1) \times 10^{-6}. \tag{7.4}$$

Table 7.2: Grouping all observed lines into 12 pairs of maximum K_i sensitivity not considering their separation in wavelength space (*rightmost* column).

Line 1	Line 2	$\Delta\mu/\mu$	ΔK_i	$\Delta\lambda$ [Å]
W0Q2	L13R1	11.5×10^{-6}	0.0553	-224.9
W0R2	L13P1	-2.2×10^{-6}	0.0529	-214.6
L1P1	L10R1	-17.2×10^{-6}	0.0431	-450.7
L1R1	L10P1	33.2×10^{-6}	0.0414	-442.3
L2P3	L9R1	-43.5×10^{-6}	0.0386	-372.5
L2R3	L9P1	41.3×10^{-6}	0.0355	-357.8
W1Q2	L12R3	-0.3×10^{-6}	0.0331	-81.7
L3P3	L10R3	28.9×10^{-6}	0.0306	-338.8
L2R1	L10P3	-49.6×10^{-6}	0.0301	-362.0
L3R3	L8R1	10.0×10^{-6}	0.0268	-261.7
L3R2	L7R1	16.7×10^{-6}	0.0207	-207.5
L3P1	L8P3	4.7×10^{-6}	0.0184	-226.3

8 Error Analysis II

8.1 Impact of wavelength calibration issues

The analysis of the line pairs in the previous section can be confirmed via bootstrap analysis fort these 12 individual $\Delta\mu$ measurements. The corresponding histogram plot is shown in Figure 8.1. The gaussian fit to the histogram gives:

$$\Delta\mu/\mu = (3.1 \pm 7.9) \times 10^{-6}, \tag{8.1}$$

and is in very good agreement of the results of the weighted linear fit carried out on the whole sample of lines at once (see Eq. 7.2) and naturally the fit to the line pairs directly (see Eq. 7.4). This further supports the confidence in the result. Furthermore it provides an indirect test for the wavelength calibration of the data.

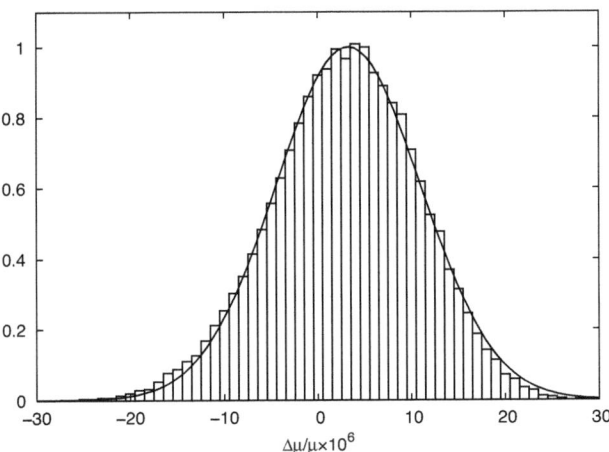

Figure 8.1: The plot shows the result of 50.000 bootstrap samples based on the average value for $\Delta\mu/\mu$ derived from the 12 lines pairs in Table 7.2. The gaussian fit yields $\Delta\mu/\mu = (3.1\pm7.9) \times 10^{-6}$ which matches the weighted fit of the measured redshifts in Equation 7.2.

The positioning error of the fit as plotted in Figure 7.1 on page 77 is not influenced by possible calibration errors since it is of statistical nature alone. The bootstrap analysis of that data delivers the same uncertainty as the unweighted fit which implies that the goodness-of-fit is reasonable and that the scatter in the data can be attributed to the positioning errors for the most part. The individual line pairs, however, are very error-prone to wavelength calibration. They cover ranges of $80 - 450$ Å. Their scatter as shown in Figure 7.4 and determined via bootstrap (Figure 8.1) yield the same error estimation for μ as the weighted fit. This implies that the statistical positioning error is the dominant source of uncertainties and that no systematics due to wavelength calibration are prominent in the data.

8.2 Test for correlation of redshift and photon energy

The results are tested for correlations of the observed redshifts to the photon energy or the excitation levels of the molecular hydrogen as described in section 5.6. Figure 8.2a reveals no present correlation between the measured redshift or the radial velocity, respectively, and the vibrational excitation of the observed transition. A detected gradient would indicate either problems with the laboratory wavelength which in fact are known to become less precise for high vibrational excitation, or some inhomogeneity in temperature or velocity of the different excitation states of the absorber along the line of sight. There is also no relation to the photon

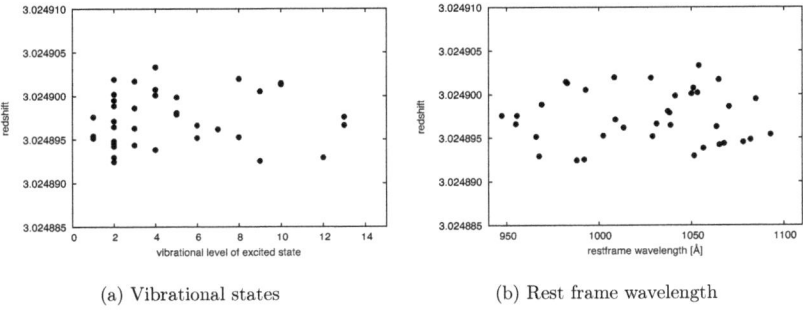

(a) Vibrational states   (b) Rest frame wavelength

Figure 8.2: Observed redshift vs. the observed vibrational states (*left*), and plotted against the rest frame wavelength directly (*right*).

energy directly as can be seen in Figure 8.2b. The shown plots have the same scale as the correlation-tests of the 2002 data. The more stringent confinement of possible correlations is evident. The absence of any such gradient support the assumption that the observed transitions of H_2 occur from the same absorber.

This is an important advantage of measuring μ compared to α for example, that -depending on the methods involved- often rely on transitions of different species that can be spatially separated or bear differential intrinsic radial velocities.

8.3 Variability of QSO 0347-383

Even though the stand-alone analysis of the data is not dependent on potential variability of the spectrum of QSO 0347-383, the constancy of observed flux should be verified before comparisons between the data of 2002 and 2009 are made. Numercial simulations of H_2 in damped Ly-α systems indicated that molecular clouds are distributed very inhomogeneously and compact in size. Hirashita et al. (2003) mention typical clump-sizes of molecule-rich regions of merely a few parsecs. Transversal velocities of these H_2 clouds may lead to different absorption features in the space of time of seven years between the data of chapter **Analysis I** and **Analysis II**.

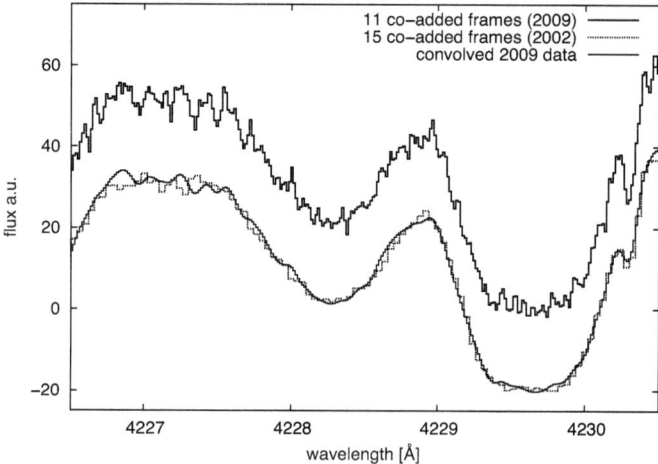

Figure 8.3: Test for variance between the recorded spectra in 2002 and 2009. The *top solid* plot graphs the 11 co-added frames of the 2009 data, whereas the *dotted* histeps represent the corresponding data of 2002 with an vertical offset for illustration. The *bottom solid* graph is a convoluted version of the *top* graph that matches the earlier data perfectly.

The two data sets that were conjointly analyzed in chapter 5 were recorded during an interval of a few weeks. The new data were recorded 7 years later, however, and the possibility of variation in the flux of QSO 0347-383 has to be taken into account. To test the data of 2002 and 2009 for consistency, the latter was convolved by a gaussian that corresponds to the resolution of the 2002 data (~ 6 kms^{-1}). The resulting flux curve was then merely scaled to give the best match (`ConvFit`). Figure 8.3 shows an exemplary region of the recent data (*top*) and with an vertical offset the co-added frames of the 2002 data (*dashed histeps*) together with the convolved data of 2009 (*solid line*).

The convolved spectrum represent a perfect match to the 2002 data. There is no variation of flux detected between the data of 2002 and 2009. The statement refers to all regions of detected H$_2$ absorption utilized for μ measurements and does not consider metal absorption lines.

8.4 Calibration and positioning errors

The positioning errors for the ascertained line parameters as estimated by RQFit is illustrated in Figure 8.4. The gaussian gives a central value of 180 m s^{-1}, which corresponds to $\sim 3 \times 10^{-6}$ in redshift or 2.5 mÅ on average which is about 1/6 of the pixel size for the 2009 data.

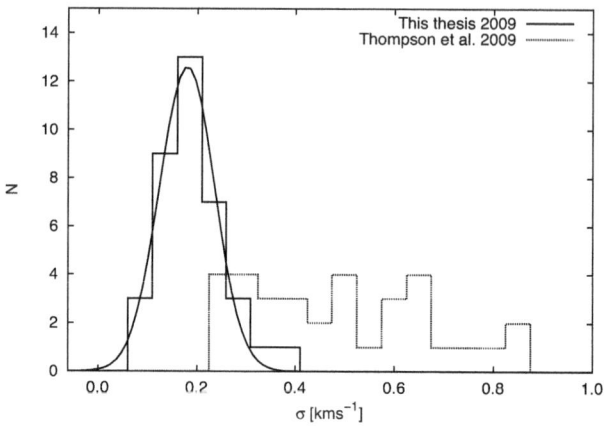

Figure 8.4: Line positioning errors in kms^{-1} for this thesis and the new data in 2009 (*solid*) and Thompson et al. (2009a) (*dotted*) based on the data of 2002, binned to 50 m s^{-1}.

A paper by Griest et al. (2010) on the wavelength accuracy of the Keck HIRES spectrograph finds inter-order offsets of up to 1000 m s^{-1}. The spectra for their analysis were taken through the Keck iodine cell which contains thousands of well calibrated iodine lines. Using these iodine exposures to calibrate the normal Th-Ar Keck data pipeline output, they found absolute wavelength offsets of 500 m s^{-1} to 1000 m s^{-1} with drifts of more than 500 m s^{-1} over a single night, and drifts of nearly 2000 m s^{-1} over several nights. These offsets correspond to an absolute redshift of uncertainty of about $\Delta \lambda \sim 2$ mÅ, with daily drifts of around $\Delta \lambda \sim 1$ mÅ and multi day drifts of nearly $\Delta \lambda \sim 4$ mÅ. Calibration uncertainties of this magnitude have an enormous impact on the measurement of fundamental constants via absorption spectra and led for example Murphy et al. (2009) to re-evaluate formerly claimed constraints.

Very recently Whitmore et al. (2010) performed a similar re-calibration of the standard UVES Th/Ar wavelength calibration pipeline using the VLT iodine cell. They find similar, but smaller, wavelength calibration errors. Offsets between successive spectra are found of the order of 100 − 400 m s^{-1} which corresponds to up to ~ 5 mÅ which is consistent with the detected shifts in section 3.2.2 for the 2002 data and section 6.2.1 for 2009. Constant velocity offsets have no direct impact on the evaluation of $\Delta \mu / \mu$. Absolute calibration is not required. Offsets from spectrum to spectrum influence the quality of co-addition or in this case of the simultaneous fit.

In particular line-widths tend to be underestimated which may lead to false error estimations. These shifts are not negligible and handled at this level of detail for the first time in this work. Properly detected, these shifts can be corrected effectively (see section 3.2.2).

Additionally, Whitmore et al. (2010) find shifts of up to 100 m s^{-1} inside the echelle orders, roughly following a saw-tooth path with its maximum near the center of each order. The error-distribution of the UVES pipeline data apparently cannot be modeled very well (see Whitmore et al. 2010). The data used in this thesis was manually reduced (see section 6.1.2 and Wendt and Molaro (2010)), and the mentioned, periodic inter-order offsets cannot be observed. The statistical positioning error of the data explains the scatter in the data points almost completely. Figure 8.5 illustrates that the error bars of the 2009 data (*filled circles*) leave little room for systematic offsets as plotted in the *solid* saw-tooth shaped graph. The *open circles* correspond to the measured redshifts in the 2002 data and are given for completeness.

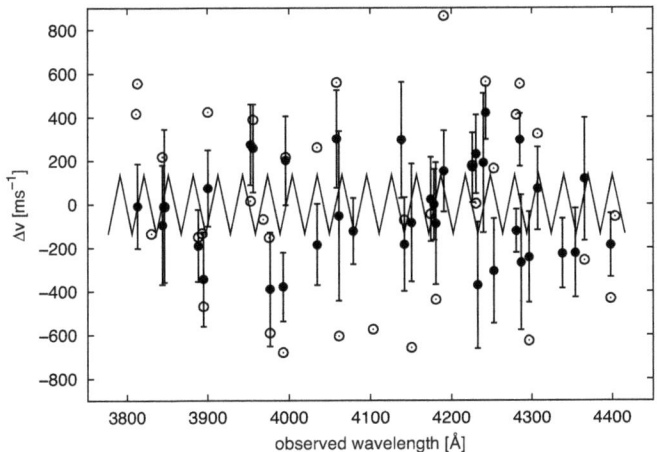

Figure 8.5: Line deviations from average redshift z_{abs} in m s^{-1} for the analysis of 2009 data (*filled circles*), the 2002 data (*open circles*) and the inter-order offsets for the VLT/UVES predicted by Whitmore et al. (2010)

The data as plotted in Figure 8.5 revealed no strong systematics or any conspicuous correlation to the saw-tooth pattern. To further assess these proclaimed inter-order calibration issues, the solar spectrum was recorded with the UVES spectrograph. To achieve such a spectrum it was required to observe Iris, a faint asteroid with an apparent magnitude of 10.1 during the observation on September 22nd, 2009.

Figure 8.6 shows the comparison of a solar spectrum taken with UVES and solar line positions as measured with HARPS. The region of about 100 Å corresponds to about 3 echelle orders for which accurate solar line positions were obtained with the HARPS spectrograph. The gap in lines around 4040 Å is due to the presence of strong Balmer lines in the solar spectrum so absorption lines are on top of large wings. The comparison does not show the Whitmore et al. (2010) modulation which is sketched in the plot. The errors in the solar lines are of < 0.6 mÅ or 45 m s^{-1}, ThAr residuals are of 35 m s^{-1} and the new measurements in the UVES spectrum are estimated to be ∼ 50 m s^{-1}, but likely less. The total error of both measurements is shown as errorbars. It seems rather unlikely that HARPS has a similar behaviour so that the two

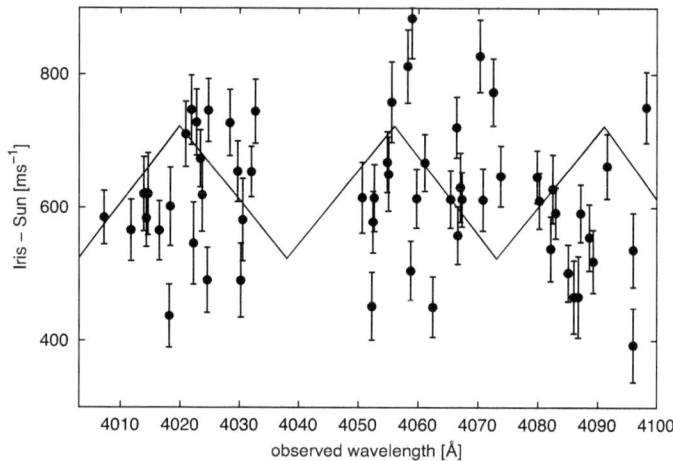

Figure 8.6: Comparison of the solar spectrum recorded with HARPS and off the Iris-Asteroid via the UVES spectrograph. The offsets show no correlation to the inter-order saw-tooth pattern (*solid lines*) described by Whitmore et al. (2010). The errorbars correspond to the total error of both measurements.

spectra cancel out the saw-tooth modulation. This is because the HARPS spectrograph has been tested with a high-precision lasercomb setup at least in one order (Wilken et al. 2010) and it does not show similar behavior with comparison to the FTS spectrum (Kurucz 2005), as confirmed by Paolo Molaro.

9 Conclusions

9.1 Inference on cosmology

The robust analysis of Chapter 3 yields

$$\Delta\mu/\mu = \left(15 \pm (9_{\text{stat}} + 6_{\text{sys}})\right) \times 10^{-6}. \tag{9.1}$$

The systematic contribution to the total uncertainty was deduced from the assessment of wavelength calibration and the data distribution itself as described in chapter 5. The systematic component that in large parts contributes to the overall error-budget of aged data was widely underestimated in several previous works and likely led to premature positive detections of variation in μ.

The analysis of data that was taken directly for the purpose of determination of changes in fundamental physical constants shows a very different quota of systematics. The details of data analysis and reduction are elucidated in chapter 6 and section 6.1.2, respectively.

The different methods to determine $\Delta\mu$ yield:

$$\Delta\mu/\mu = (7.3 \pm 9) \times 10^{-6}, \tag{9.2}$$

and

$$\Delta\mu/\mu = (7.4 \pm 7.3) \times 10^{-6}, \tag{9.3}$$

for the unweighted fits, that neglect the error estimates of the fitting procedure and merely consider the final distribution of obtained data.

The results for the weighted fit and the analysis based on single line pairs and a bootstrap analysis of the latter translate to:

$$\Delta\mu/\mu = (2.8 \pm 8.3) \times 10^{-6}, \tag{9.4}$$

$$\Delta\mu/\mu = (2.8 \pm 8.1) \times 10^{-6}, \tag{9.5}$$

$$\Delta\mu/\mu = (3.1 \pm 7.9) \times 10^{-6}, \tag{9.6}$$

respectively.

Considering the uncertainties in the line positioning of statistical nature that are in the order of 180 m s^{-1}, and the wavelength calibration residuals that are conservatively estimated to be ~ 50 m s^{-1}, the final result can be given as:

$$\Delta\mu/\mu = \left(2.9 \pm (6_{\text{stat}} + 2_{\text{sys}})\right) \times 10^{-6}. \tag{9.7}$$

The conclusion, based on the combination of two data sets from QSO 0347-383 in 2002 and recent data of 2009, is that there has been no change in the value of μ to less than 1 part in 10^5 over a time span of 11.5 Gyr. This is approximately 80% of the age of the universe. The accuracy of the limit on $\Delta\mu/\mu$ is mainly set by both the spectral resolution and the signal-to-noise ratio of the flux. This conclusion is consistent with the results of King et al. (2008) and Wendt and Reimers (2008) but inconsistent with the results of Reinhold et al. (2006). The systematic errors induced by wavelength calibration does not seem to influence the results of analysis for the 2009 data. Section 8.4 and 8.1 on page 83 show no indication of dominating systematics. For the analysis of the 2009 data, the estimated errors are consistent with the distribution of the data points. Even though systematics contributed to about 50% of the given error for the 2002 data, the dedicated observations in 2009 and their careful reduction lower the influence of systematics considerably and render the statistical portion dominant again for the current level of resolution and signal-to-noise.

What implications does a limit of $\Delta\mu/\mu < 10^{-5}$ have on theories of dark energy that invoke a rolling scalar field potential as the source of the dark energy? Chongchitnan and Efstathiou (2007) had quite some difficulties distinguishing between a universe with a cosmological constant relative to a universe with a quintessence rolling scalar field, however, the former predicts no change in μ while the latter predicts a change even though the magnitude or even the sign of the change is not presently calculable. Detection of a change in μ or its companion the fine structure constant α would be strong evidence for quintessence as opposed to a cosmological constant.

The results do, however, rule out Model A by Avelino et al. (2006) which predicts a value of $\Delta\mu/\mu = 3 \times 10^{-5}$ at a redshift of 3. This means that even at the current level of accuracy significant bounds on the quintessence models are being established. It must be pointed out though, that the theory was partially designed to match the findings of Reinhold et al. (2006), according to the main author.

Varshalovich and Potekhin (1996) note that an increase in proton mass m_p by only 0.08% would already lead to a merger of protons and electrons to form neutrons and neutrinos, whereas the reverse process –neutron beta-decay– would become energetically unfavored. By comparing the predictions of the standard model of primordial nucleosynthesis with observational data on the relative ^{4}He abundance at the current epoch, Kolb et al. (1986) concluded that the mass difference $(m_n$-$m_p)$ at the nucleosynthesis epoch corresponding to $z \sim 10^9$ did not differ from the present value by more than a few percent. However, this consideration depends both on the assumptions underlying the cosmological and primordial-nucleosynthesis models and on the values of a number of physical parameters that are not known very well. In general it should also be noted that since the law of possible variations in the fundamental constants is not known in advance, and different relations are theoretically possible (Marciano 1984), it is desirable to obtain similar constraints for different z.

In a hypothetical model by Flambaum (2006), the variations of quark mass m_q and electron masses m_e (related to variation of the Higgs vacuum field which generates fundamental masses) are given by $\delta m/m \sim 70\, \delta\alpha/\alpha$, giving an estimate of the variation for the dimensionless ratio

$$\frac{\delta(m/\Lambda_{QCD})}{(m/\Lambda_{QCD})} \sim 35\frac{\delta\alpha}{\alpha} \tag{9.8}$$

The coefficient here is model dependent but large values are generic for grand unification models in which modifications come from high energy scales; they appear because the running strong-coupling constant and Higgs constants (related to mass) run faster than α. If these models are correct, the variation in electron or quark masses and the strong interaction scale may be easier to detect than a variation in α.

According to that and further models (see chapter 2.1), a null-result in $\Delta\mu$ puts an even tighter constraint on α. This renders the estimation of the remaining uncertainty all the more important.

9.2 Outlook

A significant positive result cannot be verified by the analysis in this thesis. In the near future laboratory experiments of extraordinary precision are expected by the use of frequency comb lasers. For this new principle the 2005 Nobel Prize was awarded to John Hall and Ted Hansch. A team at the Laser Centre Vrije Universite (LCVU) is currently involved in a project that among other things, aims for $\Delta\mu/\mu$ measurements that may soon reach the disputed precision. Kajita and Moriwaki (2009) proposed to measure possible variations in μ by measuring a pure vibrational transition frequency of a cold CaH$^+$ ion. The shifts of these transitions are dominated by the Stark effect induced by the probe laser light. The uncertainty of this frequency were carefully estimated to be of the order of 10^{-16}, because the uncertainties of the Zeeman and Stark shifts are lower than that and the electric quadrupole shift is zero. The *antiproton*-to-electron mass is already measured by Barna et al. (2009) to $m_{\bar{p}}/m_e = 1836.152674(5)$, with follow-up experiments announced. The modernity of the question of variation in fundamental physical constants motivated several comparable proposals.

New data of DLAs with H$_2$ absorption features could be used to better determine different effects and systematic errors. In general a higher resolution is desirable to achieve a larger degree of confidence of the fitted parameters, mainly the broadening parameter. A better knowledge of the line shapes and more possible blends with lines that are possibly too weak to be tracked down via curve of growth analysis would increase the accuracy of the fitted line positions.

As the comparison of the 15 separate spectra in chapter 4 and of the new data in 2009 illustrates, the scatter in derived positions decreases significantly with higher resolution and in particular assiduous wavelength calibration with regard to the problem at hand. The need for data with high S/N-ratios and high resolution for the task of detection of possible μ variation, or rather finding more stringent constraints is abundantly clear. High precision in the reduction of the data is essential and probably a worthwhile subject for further studies. Recent concerns about data reduction pipelines at the KECK/Hires and VLT/UVES telescope (see section 6.1.2) and deviations from one manual reduction to another need to be faced and examined.

The majority of theories behind variations of fundamental constants is still nonspecific. Possible variations of the fine structure constant $\Delta\alpha/\alpha$ and $\Delta\mu/\mu$ cannot be pulled together and must be observed independently, despite recent tendencies to define certain correlations. A variation of the order of 2×10^{-15} yr^{-1} is in the range of modern experiments in quantum optics. However, the results obtained under local conditions of laboratory experiments cannot be simply adopted to the universe even in case of a linear behavior of variations in time. The so far conducted experiments are not directly aimed at $\Delta\mu/\mu$ measurements, since the methods require either the proton or electron mass to remain constant. Calmet and Fritzsch (2006) investigated possible scenarios of a change in proton mass without a variation of the electron mass, thus making $\Delta\alpha/\alpha$ and $\Delta\mu/\mu$ independent of each other. Despite the common belief of some yet unspecified correlation of α and μ this is remains a possibility.

Barrow (2005) points out that the local observations are based on a gross cosmological overdensity on the order of 10^{30} times denser than the mean density of the background universe and therefore cannot substitute observations on cosmological scales. Other constraints are gained from linearly extrapolating possible variations in for example proton mass towards the big bang nucleosynthesis. Among the most popular scenarios of decreasing asymptotic change in nuclear masses some neglect the drastic influence this would have on timescales close to the recombination phase (Barshay and Kreyerhoff 2006). A theory formulated by Barrow and Magueijo (2005) predicts $\Delta\mu/\mu \leq 10^{-9}$ in case of mere change in proton mass and thus expects a strong correlation between $\Delta\mu/\mu$ and $\Delta\alpha/\alpha$ via the electron mass.

It is evident that further observational results and constraints are required. The application of quasar absorption spectroscopy appears to be the most promising approach. A direct increase of precision is to be expected from further state-of-the-art observational data and larger telescopes, like the European Extremely Large Telescope (E-ELT) that will open up a wider range of H_2 detection with sufficient resolution (see, e.g., Molaro 2009; Pasquini et al. 2008; Petitjean et al. 2009b).

The search for suitable molecular systems in distant DLAs is continued (Petitjean et al. 2009a) and known systems are re-investigated with diligence. The analysis on hand is based on data recorded during a preparative observation run whose early findings led to the largest observation program dedicated to variation of fundamental constants, sheduled June 2010[1]. The observations target DLA systems for improved α and μ analysis. The program shows an unparalleled collaboration of all leading groups involved in the subject.

The data expected from these observations has the potential to set a new cornerstone in the assessment of variability of fundamental physical constants. Concerning the different approaches to analysis and contradictory findings, there is no Rosetta stone yet, though QSO 0347-383 brought several deficits to light.

Future observations featuring foreseeable instrumentation will provide further insights and allow to bring cosmological measurements of intergalactic absorbers into a new era. Spectra recorded via laser-comb calibrated spectrographs (such as CODEX or EXPRESSO) at large telescopes (E-ELT oder VLT, respectively) implicate new methods of data analysis as well. High resolution spectroscopy gains in imnportance for several fields. The Potsdam Echelle Polarimetric and Spectroscopic Instrument (PEPSI) designed for the Large Binocular Telescope (LBT) features resolutions uf up to 310.000 (Strassmeier et al. 2008).

An important point is the enhancement of absorption line modeling from simple symmetric Voigt-profiles to an more physical model incorporating radial velocity structures and inhomogeneities of the absorber.

---

[1] *"The UVES Large Program For Testing Fundamental Physics"*, ESO Program ID 185.A-0745(A)

Programs

KiComp is a simple script written in Python that computed sensitivity coefficients via comparison of transitions in molecular hydrogen and deuterium.

ErrMeter is a small Python code that performs statistical analysis of a selected wavelength interval and optionally scales the given errors accordingly.

ShiftCheck interpolates the flux of a given number of spectra and tests them for subpixel shifts between each other. Due to high CPU demands it was implemented in C/OpenMP[2].

SigTest performs simple computations to evaluate the significance of signals in the observed flux (Python).

CoAdd rebins an arbitrary number of spectral data sets and adds the flux weighted by the corresponding error. Written in C and used for illustrational purposes or line selection but not for final analysis.

GRADE is a Graphical Data Examination Tool written in C. It allows visual inspection of spectral data particularly with regard to the identifcation and fitting of H_2 absorption lines.

ScriptGen generates the input data scripts for the modified **RQFit** with regard to different spectra and common fitting parameters.

RQFit is basically identical to the program presented in Quast et al. (2005) with only minor adjustments to the problem at hand (C++/OpenMP).

ToPair re-sorts any table of obtained redshifts by the corresponding sensitivity towards changes in μ and computes $\Delta\mu/\mu$ via pairs of maximum diversity in sensitivity (Python).

BootStrap resamples an arbitrary number of data points for boostrap analysis and applies a gaussian fit to the resulting histogram. Written in C/OpenMP to allow for large samples.

2DimErrFit performs a linear fit to data with errors in both dimensions. For the problem at hand, it takes into account ΔK_i in addition to Δz (C).

---

[2]OpenMP (Open Multi-Processing) is an application programming interface (API) that supports multi-platform shared memory multiprocessing programming in C, C++ and Fortran on many architectures, including Unix and Microsoft Windows platforms. It consists of a set of compiler directives, library routines, and environment variables that influence run-time behavior (see, e.g., Müller et al. 2009).

SpecSim generates synthetic spectra with H_2 and $H\,\textsc{i}$ features including gaussian and poisson noise as well as pixel-to-pixel correlations (C).

ConvFit convolves a spectrum with a given gaussian to compare two spectra of different resolution with each other. The scaling factor for the best match is fitted as well to evaluate the consistency of the two data sets.

List of Figures

2.1	Refinement of restframe wavelengths	16
2.2	Sensitivity coefficients of observed lines	17
2.3	Differences between derived K coefficients	18
2.4	Measured proton-to-electron mass ratio	21
3.1	QSO 0347-383 (STScI)	23
3.2	Co-added spectrum of QSO 0347-383	24
3.3	Saturated absorption	26
3.4	Interpolated flux	28
3.5	Sub-pixel cross-correlation	29
3.6	Separate and co-added spectra	31
3.7	Absolute offsets between 52 lines in two datasets	32
3.8	Exemplary lines that fail the selection criterion	33
3.9	Redshift vs. sensitivity coefficient	34
3.10	Line positioning errors	34
3.11	Continuum matching via parabolic fit	40
4.1	Unweighted fit for QSO 0347-383	42
4.2	Average positioning error	43
4.3	Result based on individual line pairs	46
5.1	Bootstrap analysis	50
5.2	Influence of initial shift correction	54
5.3	Rotational states	55
5.4	Rotational states - bootstrap	56
5.5	Test for redshift - photon energy correlation	57
5.6	Lyman and Werner band transitions	59
5.7	Bootstrap analysis of Lyman lines only	60
5.8	Simulation of fits	61
5.9	Mean error of fits of synthetic spectra	62
5.10	Net shifts of fits of synthetic spectra	63
6.1	Example region of co-added 2002 and 2009 data	66
6.2	Separate and co-added spectra	67

6.3	Subsampling of flux via	71
6.4	Lineselection via GRADE	73
6.5	Example region of co-added 2002 and 2009 data	74
6.6	Additional components for the fit	75
6.7	Significance test for absorption features	76
7.1	Unweighted fit for QSO 0347-383	77
7.2	Redshift vs. sensitivity coefficient	78
7.3	Bootstrap analysis 2009	80
7.4	Result based on individual line pairs	81
8.1	Bootstrap analysis based on individual line pairs	83
8.2	Test for redshift - photon energy correlation	85
8.3	Variance test between 2002 and 2009 data.	86
8.4	Line positioning errors 2009	88
8.5	Line positioning deviations	89
8.6	UVES calibration	90

List of Tables

2.1	DLAs with H_2 absorption	14
3.1	Journal of the observations	24
3.2	Relative shifts of the observed spectra	30
3.3	Excluded lines	35
4.1	QSO 0347-383 Line List	45
4.2	Line pairs of maximum K_i sensitivity	48
6.1	Journal of the observations (2009 data)	65
6.2	Relative shifts of the observed spectra (2009 data)	72
7.1	QSO 0347-383 Line List (2009 data)	79
7.2	Line pairs of maximum K_i sensitivity (2009 data)	82

Acknowledgements

I want to express my gratitude towards Prof. Dieter Reimers for giving me the unique opportunity to work in this vivid field. Also I am indebted to Dr. Robert Baade for the careful inspection of this work and numerous discussions on the universe, its physics and its habitants. In addition I thank David Simmons for proof-reading on short notice.

And of course I am thankful for my life partner Claudia and my parents for backing me all the time throughout this endeavor.

> "The universe is not required
> to be in perfect harmony
> with human ambition."
> - Carl Sagan

References

H. Abgrall, E. Roueff, F. Launay, J. Y. Roncin, and J. L. Subtil. Table of the Lyman Band System of Molecular Hydrogen. *A&AS*, 101:273–+, October 1993a.

H. Abgrall, E. Roueff, F. Launay, J. Y. Roncin, and J. L. Subtil. Table of the Werner Band System of Molecular Hydrogen. *A&AS*, 101:323–+, October 1993b.

H. Abgrall, E. Roueff, and I. Drira. Total transition probability and spontaneous radiative dissociation of B, C, B' and D states of molecular hydrogen. *A&AS*, 141:297–300, January 2000. doi: 10.1051/aas:2000121.

A.C. Aitken. On least squares and linear combinations of observations. *Proc.Roy.Soc.Edinburgh*, 55:42–48, 1934.

R. D. Atkinson. Secular Changes in Atomic "Constants". *Physical Review*, 170:1193–1194, June 1968. doi: 10.1103/PhysRev.170.1193.

P. P. Avelino, C. J. A. P. Martins, N. J. Nunes, and K. A. Olive. Reconstructing the dark energy equation of state with varying couplings. *Phys.Rev.*, 74(8):083508–+, October 2006. doi: 10.1103/PhysRevD.74.083508.

D. Bailly, E. J. Salumbides, M. Vervloet, and W. Ubachs. Accurate level energies in the states of H2. *Molecular Physics*, 2010. doi: doi:10.1080/00268970903413350.

D. Barna, A. Dax, J. Eades, K. Gomikawa, R. S. Hayano, M. Hori, D. Horváth, B. Juhász, N. Ono, W. Pirkl, E. Widmann, and H. A. Torii. Determination of the antiproton-to-electron mass ratio by laser spectroscopy. *Hyperfine Interactions*, 194:1–6, November 2009. doi: 10.1007/s10751-009-0022-9.

J. Barrow. Varying constants. *Royal Society of London Philosophical Transactions Series A*, 363:2139–2153, September 2005.

J. D. Barrow and J. Magueijo. Cosmological constraints on a dynamical electron mass. *PhysRev*, 72(4):043521–+, August 2005.

S. Barshay and G. Kreyerhoff. An asymptotic decrease of (m_p/m_e) with cosmological time, from a decreasing, small effective vacuum expectation value moving from a potential maximum in the early universe. *ArXiv Astrophysics e-prints*, June 2006.

S. Blatt, A. D. Ludlow, G. K. Campbell, J. W. Thomsen, T. Zelevinsky, M. M. Boyd, J. Ye, X. Baillard, M. Fouché, R. Le Targat, A. Brusch, P. Lemonde, M. Takamoto, F.-L. Hong, H. Katori, and V. V. Flambaum. New Limits on Coupling of Fundamental Constants to Gravity Using Sr87 Optical Lattice Clocks. *Physical Review Letters*, 100(14):140801–+, April 2008. doi: 10.1103/PhysRevLett.100.140801.

R. W. Boyd and C. Braun. Nonlinear Optics. *Applied Optics*, 31:6583–6584, November 1992.

R. N. Cahn. The eighteen arbitrary parameters of the standard model in your everyday life. *Reviews of Modern Physics*, 68:951–959, July 1996. doi: 10.1103/RevModPhys.68.951.

X. Calmet and H. Fritzsch. A time variation of proton-electron mass ratio and grand unification. *Europhysics Letters*, 76:1064–1067, December 2006. doi: 10.1209/epl/i2006-10393-0.

G. R. Carruthers. Rocket Observation of Interstellar Molecular Hydrogen. *ApJL*, 161:L81+, August 1970. doi: 10.1086/180575.

M. Centurión, P. Molaro, and S. Levshakov. Calibration issues in estimating variability of the fine structure constant (alpha) with cosmic time. *ArXiv e-prints*, October 2009.

T. Chiba, T. Kobayashi, M. Yamaguchi, and J. Yokoyama. Time variation of the proton-electron mass ratio and the fine structure constant with a runaway dilaton. *Phys.Rev.*, 75 (4):043516–+, February 2007. doi: 10.1103/PhysRevD.75.043516.

S. Chongchitnan and G. Efstathiou. Can we ever distinguish between quintessence and a cosmological constant? *Phys.Rev.*, 76(4):043508–+, August 2007. doi: 10.1103/PhysRevD.76.043508.

F. Combes and G. Pineau Des Forets. Molecular Hydrogen in Space. November 2000.

L. L. Cowie and A. Songaila. Astrophysical Limits on the Evolution of Dimensionless Physical Constants over Cosmological Time. *ApJ*, 453:596–+, November 1995. doi: 10.1086/176422.

T. E. Cravens. Vibrationally excited molecular hydrogen in the upper atmosphere of Jupiter. *Journal of Geophysical Research*, 92:11083–11100, October 1987. doi: 10.1029/JA092iA10p11083.

J. Cui, J. Bechtold, J. Ge, and D. M. Meyer. Molecular Hydrogen in the Damped Lyα Absorber of Q1331+170. *ApJ*, 633:649–663, November 2005. doi: 10.1086/444368.

T. Damour. Questioning the Equivalence Principle. *ArXiv General Relativity and Quantum Cosmology e-prints*, September 2001.

M. Dine, Y. Nir, G. Raz, and T. Volansky. Time variations in the scale of grand unification. *Phys.Rev.*, 67(1):015009–+, January 2003. doi: 10.1103/PhysRevD.67.015009.

P. A. M. Dirac. The Cosmological Constants. *Nature*, 139:323–+, February 1937. doi: 10.1038/139323a0.

S. D'Odorico, M. Dessauges-Zavadsky, and P. Molaro. A new deuterium abundance measurement from a damped Ly-alpha system at z_{abs} =3.025. *A&A*, 368:L21–L24, March 2001. doi: 10.1051/0004-6361:20010183.

J. L. Dunham. The Energy Levels of a Rotating Vibrator. *Physical Review*, 41:721–731, September 1932.

B. Edlén. The Refractive Index of Air. *Metrologia*, 2:71–80, April 1966. doi: 10.1088/0026-1394/2/2/002.

B. Efron and R. J Tibshirani. *An Introduction to the Bootstrap*. 1986.

D. Field, M. Gerin, S. Leach, J. L. Lemaire, G. Pineau Des Forets, F. Rostas, D. Rouan, and D. Simons. High spatial resolution observations of H_2 vibrational fluorescence in NGC 2023. *A&A*, 286:909–914, June 1994.

V. V. Flambaum. Variation of Fundamental Constants. In C. Roos, H. Haffner, & R. Blatt, editor, *Atomic Physics 20*, volume 869 of *American Institute of Physics Conference Series*, pages 29–36, November 2006. doi: 10.1063/1.2400630.

J. L. Flowers and B. W. Petley. Progress in our knowledge of the fundamental constants of physics. *Reports on Progress in Physics*, 64:1191–1246, October 2001. doi: 10.1088/0034-4885/64/10/201.

C. B. Foltz, F. H. Chaffee, Jr., and J. H. Black. Molecules at early epochs. IV - Confirmation of the detection of H2 toward PKS 0528 - 250. *ApJ*, 324:267–278, January 1988. doi: 10.1086/165893.

H. Fritzsch. The fundamental constants in physics. *Physics-Uspekhi*, 52(4):359, 2009. URL http://stacks.iop.org/1063-7869/52/i=4/a=A04.

G. Gamow. Electricity, Gravity, and Cosmology. *Physical Review Letters*, 19:759–761, September 1967. doi: 10.1103/PhysRevLett.19.759.

J. Ge and J. Bechtold. Molecular Hydrogen Absorption in the z = 1.97 Damped Ly alpha Absorption System toward Quasi-stellar Object Q0013-004. *ApJL*, 477:L73+, March 1997. doi: 10.1086/310527.

K. Griest, J. B. Whitmore, A. M. Wolfe, J. X. Prochaska, J. C. Howk, and G. W. Marcy. Wavelength Accuracy of the Keck HIRES Spectrograph and Measuring Changes in the Fine Structure Constant. *ApJ*, 708:158–170, January 2010. doi: 10.1088/0004-637X/708/1/158.

E. Habart, F. Boulanger, L. Verstraete, C. M. Walmsley, and G. Pineau des Forêts. Some empirical estimates of the H_2 formation rate in photon-dominated regions. *A&A*, 414:531–544, February 2004. doi: 10.1051/0004-6361:20031659.

Nikolaus Hansen and Andreas Ostermeier. Completely derandomized self-adaptation in evolution strategies. *Evolutionary Computation*, 9:159–195, 2001.

G. Herzberg. Quadrupole Rotation-Vibration Spectrum of the Hydrogen Molecule. *Nature*, 163:170–+, January 1949. doi: 10.1038/163170a0.

H. Hirashita, A. Ferrara, K. Wada, and P. Richter. Molecular hydrogen in damped Lyα systems: spatial distribution. *MNRAS*, 341:L18–L22, May 2003. doi: 10.1046/j.1365-8711.2003.06648.x.

C. J. Hogan. Why the universe is just so. *Reviews of Modern Physics*, 72:1149–1161, October 2000. doi: 10.1103/RevModPhys.72.1149.

U. Hollenstein, E. Reinhold, C. A. de Lange, and W. Ubachs. LETTER TO THE EDITOR: High-resolution VUV-laser spectroscopic study of the Lyman bands in H_2 and HD. *Journal of Physics B Atomic Molecular Physics*, 39:L195–L201, April 2006. doi: 10.1088/0953-4075/39/8/L02.

A. Ivanchik, P. Petitjean, E. Rodriguez, and D. Varshalovich. Does the proton-to-electron mass ratio $\mu = m_p/m_e$ vary in the course of cosmological evolution? *Astrophysics and Space Science*, 283:583–588, 2003. doi: 10.1023/A:1022522600509.

A. Ivanchik, P. Petitjean, D. Varshalovich, B. Aracil, R. Srianand, H. Chand, C. Ledoux, and P. Boissé. A new constraint on the time dependence of the proton-to-electron mass ratio. Analysis of the Q 0347-383 and Q 0405-443 spectra. *A&A*, 440:45–52, September 2005. doi: 10.1051/0004-6361:20052648.

A. V. Ivanchik, E. Rodriguez, P. Petitjean, and D. A. Varshalovich. Do the Fundamental Constants Vary in the Course of Cosmological Evolution? *Astronomy Letters*, 28:423–427, July 2002. doi: 10.1134/1.1491963.

A.V. Ivanchik, E. Rodriguez, P. Petitjean, and D. Varshalovich. Do the Fundamental Constants Vary in the Course of the Cosmological Evolution? 2001.

T. I. Ivanov, M. O. Vieitez, C. A. de Lange, and W. Ubachs. Frequency calibration of $B^1\Sigma_u^+$ X $^1\Sigma_g^+$ (6,0) Lyman transitions in H_2 for comparison with quasar data. *Journal of Physics B Atomic Molecular Physics*, 41(3):035702–+, February 2008. doi: 10.1088/0953-4075/41/3/035702.

D. E. Jennings, S. L. Bragg, and J. W. Brault. The V = 0 - 0 spectrum of H2. *ApJL*, 282: L85–L88, July 1984.

M. Kajita and Y. Moriwaki. Proposed detection of variation in m_p/m_e using a vibrational transition frequency of a CaH^+ ion. *Journal of Physics B Atomic Molecular Physics*, 42(15): 154022–+, August 2009. doi: 10.1088/0953-4075/42/15/154022.

A. Kaufer, S. DâĂŹOdorico, L. Kaper, C. Ledoux, G. James, H. Sana, and J. Smoker. *Very Large Telescope Paranal Science Operations UV-Visual Echelle Spectrograph User manual.* 2004.

J. A. King, J. K. Webb, M. T. Murphy, and R. F. Carswell. Stringent Null Constraint on Cosmological Evolution of the Proton-to-Electron Mass Ratio. *Physical Review Letters*, 101 (25):251304–+, December 2008. doi: 10.1103/PhysRevLett.101.251304.

E. W. Kolb, M. J. Perry, and T. P. Walker. Time variation of fundamental constants, primordial nucleosynthesis, and the size of extra dimensions. *Physical Review*, 33:869–871, February 1986. doi: 10.1103/PhysRevD.33.869.

R. L. Kurucz. New atlases for solar flux, irradiance, central intensity, and limb intensity. *Memorie della Societa Astronomica Italiana Supplement*, 8:189–+, 2005.

C. Ledoux, R. Srianand, and P. Petitjean. Detection of molecular hydrogen in a near Solar-metallicity damped Lyman-alpha system at $z_{abs} \approx 2$ toward Q 0551-366. *A&A*, 392:781–789, September 2002. doi: 10.1051/0004-6361:20021187.

C. Ledoux, P. Petitjean, and R. Srianand. The Very Large Telescope Ultraviolet and Visible Echelle Spectrograph survey for molecular hydrogen in high-redshift damped Lyman α systems. *MNRAS*, 346:209–228, November 2003. doi: 10.1046/j.1365-2966.2003.07082.x.

C. Ledoux, P. Petitjean, and R. Srianand. Molecular Hydrogen in a Damped Lyα System at z_{abs}=4.224. *ApJL*, 640:L25–L28, March 2006. doi: 10.1086/503278.

S. A. Levshakov and D. A. Varshalovich. Molecular hydrogen in the Z = 2.811 absorbing material toward the quasar PKS 0528-250. *MNRAS*, 212:517–521, February 1985.

S. A. Levshakov, P. Molaro, M. Centurión, S. D'Odorico, P. Bonifacio, and G. Vladilo. UVES observations of QSO 0000-2620: molecular hydrogen abundance in the damped Lyalpha system at $z_{abs} = 3.3901$. *A&A*, 361:803–810, September 2000.

S. A. Levshakov, M. Dessauges-Zavadsky, S. D'Odorico, and P. Molaro. A new constraint on cosmological variability of the proton-to-electron mass ratio. *MNRAS*, 333:373–377, June 2002. doi: 10.1046/j.1365-8711.2002.05408.x.

J.-M. Lévy-Leblond. On the conceptual nature of the physical constants. *Nuovo Cimento Rivista Serie*, 7:187–214, April 1977. doi: 10.1007/BF02748049.

A. L. Malec, R. Buning, M. T. Murphy, N. Milutinovic, S. L. Ellison, J. X. Prochaska, L. Kaper, J. Tumlinson, R. F. Carswell, and W. Ubachs. New limit on a varying proton-to-electron mass ratio from high-resolution optical quasar spectra. *ArXiv e-prints*, January 2010.

D. Maoz, J. N. Bahcall, D. P. Schneider, N. A. Bahcall, S. Djorgovski, R. Doxsey, A. Gould, S. Kirhakos, G. Meylan, and B. Yanny. The Hubble Space Telescope Snapshot Survey. IV - A summary of the search for gravitationally lensed quasars. *ApJ*, 409:28–41, May 1993. doi: 10.1086/172639.

W. J. Marciano. Time variation of the fundamental 'constants' and Kaluza-Klein theories. *Physical Review Letters*, 52:489–491, February 1984. doi: 10.1103/PhysRevLett.52.489.

V. V. Meshkov, A. V. Stolyarov, A. V. Ivanchik, and D. A. Varshalovich. Ab initio nonadiabatic calculation of the sensitivity coefficients for lines of H2 to the proton-to-electron mass ratio. *Soviet Journal of Experimental and Theoretical Physics Letters*, 83:303–307, June 2006. doi: 10.1134/S0021364006080017.

P. J. Mohr and B. N. Taylor. CODATA recommended values of the fundamental physical constants: 2002. *Reviews of Modern Physics*, 77:1–107, January 2005. doi: 10.1103/RevModPhys.77.1.

P. Molaro. New spectrographs for the VLT and E-ELT suited for the measurement of fundamental constants variability . *Memorie della Societa Astronomica Italiana*, 80:912–+, 2009.

P. Molaro, S. A. Levshakov, S. Monai, M. Centurión, P. Bonifacio, S. D'Odorico, and L. Monaco. UVES radial velocity accuracy from asteroid observations. I. Implications for fine structure constant variability. *A&A*, 481:559–569, April 2008a. doi: 10.1051/0004-6361:20078864.

P. Molaro, D. Reimers, I. I. Agafonova, and S. A. Levshakov. Bounds on the fine structure constant variability from Fe ii absorption lines in QSO spectra. *European Physical Journal Special Topics*, 163:173–189, October 2008b. doi: 10.1140/epjst/e2008-00818-4.

H. W. Moos, W. C. Cash, L. L. Cowie, and et al. Davidsen. Overview of the Far Ultraviolet Spectroscopic Explorer Mission. *ApJL*, 538:L1–L6, July 2000. doi: 10.1086/312795.

D. C. Morton and H. L. Dinerstein. Interstellar molecular hydrogen toward zeta Puppis. *ApJ*, 204:1–11, February 1976. doi: 10.1086/154144.

D. C. Morton, A. E. Wright, B. A. Peterson, D. L. Jauncey, and J. Chen. Absorption lines and ion abundances in the QSO PKS 0528-250. *MNRAS*, 193:399–413, November 1980.

M. S. Müller, B. R. de Supinski, and B. M. Chapman. Evolving OpenMP in an Age of Extreme Parallelism. *Lecture Notes in Computer Science*, 5568, 2009. doi: 10.1007/978-3-642-02303-3.

M. T. Murphy, J. K. Webb, and V. V. Flambaum. Further evidence for a variable fine-structure constant from Keck/HIRES QSO absorption spectra. *MNRAS*, 345:609–638, October 2003. doi: 10.1046/j.1365-8711.2003.06970.x.

M. T. Murphy, J. K. Webb, and V. V. Flambaum. Revision of VLT/UVES constraints on a varying fine-structure constant. *MNRAS*, 384:1053–1062, March 2008. doi: 10.1111/j.1365-2966.2007.12695.x.

M. T. Murphy, J. K. Webb, and V. V. Flambaum. Keck constraints on a varying fine-structure constant: wavelength calibration errors . *Memorie della Societa Astronomica Italiana*, 80: 833–+, 2009.

P. Noterdaeme, C. Ledoux, P. Petitjean, and R. Srianand. Molecular hydrogen in high-redshift damped Lyman-α systems: the VLT/UVES database. *A&A*, 481:327–336, April 2008. doi: 10.1051/0004-6361:20078780.

P. S. Osmer and M. G. Smith. Discovery and spectrophotometry of the quasars in the -40 deg zone of the CTIO Curtis Schmidt survey. *ApJS*, 42:333–349, February 1980. doi: 10.1086/190653.

B. E. J. Pagel. Implications of quasar spectroscopy for constancy of constants. *Royal Society of London Philosophical Transactions Series A*, 310:245–247, December 1983.

L. Pasquini, G. Avila, H. Dekker, B. Delabre, S. D'Odorico, A. Manescau, M. Haehnelt, B. Carswell, R. Garcia-Lopez, R. Lopez, M. T. Osorio, R. Rebolo, S. Cristiani, P. Bonifacio, V. D'Odorico, P. Molaro, P. Spanò, F. Zerbi, M. Mayor, M. Dessauges, D. Megevand, F. Pepe, D. Queloz, and S. Udry. CODEX: the high-resolution visual spectrograph for the E-ELT. In *Society of Photo-Optical Instrumentation Engineers (SPIE) Conference Series*, volume 7014 of *Society of Photo-Optical Instrumentation Engineers (SPIE) Conference Series*, August 2008. doi: 10.1117/12.787936.

P. Petitjean, R. Srianand, and C. Ledoux. Molecular hydrogen at z_{abs} =1.973 toward Q0013-004: dust depletion pattern in damped Lyman α systems. *MNRAS*, 332:383–391, May 2002. doi: 10.1046/j.1365-8711.2002.05331.x.

P. Petitjean, C. Ledoux, P. Noterdaeme, and R. Srianand. Metallicity as a criterion to select H_2-bearing damped Lyman-α systems. *A&A*, 456:L9–L12, September 2006. doi: 10.1051/0004-6361:20065769.

P. Petitjean, P. Noterdaeme, R. Srianand, C. Ledoux, A. Ivanchik, and N. Gupta. Searching for places where to test the variations of fundamental constants. *Memorie della Societa Astronomica Italiana*, 80:859–+, 2009a.

P. Petitjean, R. Srianand, H. Chand, A. Ivanchik, P. Noterdaeme, and N. Gupta. Constraining Fundamental Constants of Physics with Quasar Absorption Line Systems. *Space Science Reviews*, 148:289–300, December 2009b. doi: 10.1007/s11214-009-9520-y.

B. W. Petley. New definition of the metre. *Nature*, 303:373–376, June 1983. doi: 10.1038/303373a0.

J. Philip, J.P. Sprengers, Th. Pielage, C.A. Lange, W. Ubachs, and E. Reinhold. highly accurate transition frequencies in the H2 Lyman and Werner absorption bands. *Can. J. Chem*, 82: 712–722, 2004.

A. Y. Potekhin, A. V. Ivanchik, D. A. Varshalovich, K. M. Lanzetta, J. A. Baldwin, G. M. Williger, and R. F. Carswell. Testing Cosmological Variability of the Proton-to-Electron Mass Ratio Using the Spectrum of PKS 0528-250. *ApJ*, 505:523–528, October 1998. doi: 10.1086/306211.

R. Quast, R. Baade, and D. Reimers. Evolution strategies applied to the problem of line profile decomposition in QSO spectra. *A&A*, 431:1167–1175, March 2005. doi: 10.1051/0004-6361:20041601.

D. Reimers, R. Baade, R. Quast, and S. A. Levshakov. Detection of molecular hydrogen at z = 1.15 toward HE 0515-4414. *A&A*, 410:785–793, November 2003. doi: 10.1051/0004-6361:20031313.

E. Reinhold, R. Buning, U. Hollenstein, A. Ivanchik, P. Petitjean, and W. Ubachs. Indication of a Cosmological Variation of the Proton-Electron Mass Ratio Based on Laboratory Measurement and Reanalysis of H2 Spectra. *Physical Review Letters*, 96(15):151101–+, April 2006. doi: 10.1103/PhysRevLett.96.151101.

E. J. Salumbides, D. Bailly, A. Khramov, A. L. Wolf, K. S. E. Eikema, M. Vervloet, and W. Ubachs. Improved Laboratory Values of the H_2 Lyman and Werner Lines for Constraining Time Variation of the Proton-to-Electron Mass Ratio. *Physical Review Letters*, 101(22): 223001–+, November 2008. doi: 10.1103/PhysRevLett.101.223001.

B. D. Savage, R. C. Bohlin, J. F. Drake, and W. Budich. A survey of interstellar molecular hydrogen. I. *ApJ*, 216:291–307, August 1977. doi: 10.1086/155471.

L. Spitzer, Jr., W. D. Cochran, and A. Hirshfeld. Column densities of interstellar molecular hydrogen. *ApJS*, 28:373–389, October 1974. doi: 10.1086/190323.

R. Srianand, P. Petitjean, C. Ledoux, G. Ferland, and G. Shaw. The VLT-UVES survey for molecular hydrogen in high-redshift damped Lyman α systems: physical conditions in the neutral gas. *MNRAS*, 362:549–568, September 2005. doi: 10.1111/j.1365-2966.2005.09324.x.

K. G. Strassmeier, M. Woche, I. Ilyin, E. Popow, S.-M. Bauer, F. Dionies, T. Fechner, M. Weber, A. Hofmann, J. Storm, R. Materne, W. Bittner, J. Bartus, T. Granzer, C. Denker, T. Carroll, M. Kopf, I. DiVarano, E. Beckert, and M. Lesser. PEPSI: the Potsdam Echelle Polarimetric and Spectroscopic Instrument for the LBT. In *Society of Photo-Optical Instrumentation Engineers (SPIE) Conference Series*, volume 7014 of *Society of Photo-Optical Instrumentation Engineers (SPIE) Conference Series*, August 2008. doi: 10.1117/12.787376.

A. Stuart and J. K. Ord. *Kendall's advanced theory of statistics. Vol.1: Distribution theory*. 1994.

R. I. Thompson. The determination of the electron to proton inertial mass ratio via molecular transitions. *Astrophysical Letters*, 16:3–+, 1975.

R. I. Thompson, J. Bechtold, J. H. Black, D. Eisenstein, X. Fan, R. C. Kennicutt, C. Martins, J. X. Prochaska, and Y. L. Shirley. An Observational Determination of the Proton to Electron Mass Ratio in the Early Universe. *ApJ*, 703:1648–1662, October 2009a. doi: 10.1088/0004-637X/703/2/1648.

R. I. Thompson, J. Bechtold, J. H. Black, and C. J. A. P. Martins. Alternative data reduction procedures for UVES: Wavelength calibration and spectrum addition. *New Astronomy*, 14: 379–390, May 2009b. doi: 10.1016/j.newast.2008.11.001.

W. Ubachs, R. Buning, K. S. E. Eikema, and E. Reinhold. On a possible variation of the proton-to-electron mass ratio: H_2 spectra in the line of sight of high-redshift quasars and in the laboratory. *Journal of Molecular Spectroscopy*, 241:155–179, February 2007. doi: 10.1016/j.jms.2006.12.004.

J.-P. Uzan. The fundamental constants and their variation: observational and theoretical status. *Reviews of Modern Physics*, 75:403–455, April 2003. doi: 10.1103/RevModPhys.75.403.

D. A. Varshalovich and S. A. Levshakov. On a time dependence of physical constants. *Soviet Journal of Experimental and Theoretical Physics Letters*, 58:237–240, August 1993.

D. A. Varshalovich and A. Y. Potekhin. Cosmological Variability of Fundamental Physical Constants. *Space Science Reviews*, 74:259–268, November 1995. doi: 10.1007/BF00751411.

D. A. Varshalovich and A. Y. Potekhin. Have the masses of molecules changed during the lifetime of the Universe? *Astronomy Letters*, 22:1–5, January 1996.

D. A. Varshalovich, V. E. Panchuk, and A. V. Ivanchik. Absorption systems in the spectra of the QSOs HS 1946+76, S5 0014+81, and S4 0636+68: New constraints on cosmological variation of the fine-structure constant. *Astronomy Letters*, 22:6–13, January 1996.

J. K. Webb, V. V. Flambaum, C. W. Churchill, M. J. Drinkwater, and J. D. Barrow. Search for Time Variation of the Fine Structure Constant. *Physical Review Letters*, 82:884–887, February 1999. doi: 10.1103/PhysRevLett.82.884.

J. K. Webb, M. T. Murphy, V. V. Flambaum, V. A. Dzuba, J. D. Barrow, C. W. Churchill, J. X. Prochaska, and A. M. Wolfe. Further Evidence for Cosmological Evolution of the Fine Structure Constant. *Physical Review Letters*, 87(9):091301–+, August 2001. doi: 10.1103/PhysRevLett.87.091301.

S. Weinberg. Overview of theoretical prospects for understanding the values of fundamental constants. In *(Royal Society, Discussion on the Constants of Physics, London, England, May 25, 26, 1983) Royal Society (London), Philosophical Transactions, Series A (ISSN 0080-4614), vol. 310, no. 1512, Dec. 20, 1983, p. 249-252.*, volume 310, pages 249–252, December 1983.

M. Wendt and P. Molaro. Robust limit on a varying proton-to-electron mass ratio from a single H_2 system. *A&A submitted*, 2010.

M. Wendt and D. Reimers. Variability of the proton-to-electron mass ratio on cosmological scales. *European Physical Journal Special Topics*, 163:197–206, October 2008. doi: 10.1140/epjst/e2008-00820-x.

M. Wendt, D. Reimers, and P. Molaro. Cosmological observations to shed light on possible variations. expectations, limitations and status quo. *Memorie della Societa Astronomica Italiana*, 80:876–+, 2009.

J. B. Whitmore, M. T. Murphy, and K. Griest. Wavelength Calibration of the VLT-UVES Spectrograph. *ArXiv e-prints*, April 2010.

T. Wilken, C. Lovis, A. Manescau, T. Steinmetz, L. Pasquini, G. Lo Curto, T. W. Hänsch, R. Holzwarth, and T. Udem. High-precision calibration of spectrographs. *MNRAS*, pages L55+, April 2010. doi: 10.1111/j.1745-3933.2010.00850.x.

E. Witten. Some properties of O(32) superstrings. *Physics Letters B*, 149:351–356, December 1984. doi: 10.1016/0370-2693(84)90422-2.

I want morebooks!

Buy your books fast and straightforward online - at one of world's fastest growing online book stores! Environmentally sound due to Print-on-Demand technologies.

Buy your books online at
www.morebooks.shop

Kaufen Sie Ihre Bücher schnell und unkompliziert online – auf einer der am schnellsten wachsenden Buchhandelsplattformen weltweit! Dank Print-On-Demand umwelt- und ressourcenschonend produziert.

Bücher schneller online kaufen
www.morebooks.shop

KS OmniScriptum Publishing
Brivibas gatve 197
LV-1039 Riga, Latvia
Telefax +371 686 204 55

info@omniscriptum.com
www.omniscriptum.com

Printed by Books on Demand GmbH, Norderstedt / Germany